ALLOY PHASE EQUILIBRIA

ALLOY PHASE EQUILIBRIA

A. PRINCE

ELSEVIER PUBLISHING COMPANY
AMSTERDAM — LONDON — NEW YORK
1966

ELSEVIER PUBLISHING COMPANY
335 JAN VAN GALENSTRAAT, P.O. BOX 211, AMSTERDAM

AMERICAN ELSEVIER PUBLISHING COMPANY, INC.
52 VANDERBILT AVENUE, NEW YORK, N.Y. 10017

ELSEVIER PUBLISHING COMPANY LIMITED
RIPPLESIDE COMMERCIAL ESTATE, BARKING, ESSEX

LIBRARY OF CONGRESS CATALOG CARD NUMBER 65–13898

WITH 277 ILLUSTRATIONS AND 16 TABLES

PRINTED IN THE NETHERLANDS

Preface

A metallurgist has been defined as a person who thinks in terms of phase diagrams. Whether one accepts this statement or not, it is self-evident that an ability to interpret phase equilibria is essential to the professional metallurgist. Phase diagrams, equilibrium diagrams or constitutional diagrams, as they are variously called, are convenient methods for concisely plotting the equilibrium relationships in alloy systems. They have often been criticised for not providing the information they are not intended to yield. By their very nature they only consider thermodynamically equilibrium conditions, they give no clues as to the rates of reaction, they yield no information on the effect of point and line defects on the properties of the phases, they say nothing about the phase distribution morphologically, and they ignore surface energy effects at phase boundaries and strain energy effects in transformations. Naturally, many ancillary data on these topics are gathered in the course of the determination of a phase diagram.

The present volume deals with alloy phase equilibria only. It does not attempt to trespass on the complementary fields of the kinetics of phase transformations and only passing reference is made to the structural approach to alloying behaviour. The latter has been admirably covered by the internationally acclaimed textbooks of W. HUME-ROTHERY and G. V. RAYNOR in the Institute of Metals Monograph Series. It is taken for granted that the reader of this book has access to copies of these for reference.

The majority of previous works on phase equilibria have not dealt with the thermodynamic approach. As the work of O. KUBASCHEWSKI and others has demonstrated, the application of thermodynamics to phase equilibria can provide a useful check on experimental data and allow its extension to regions difficult to determine experimentally. In view of this the first two chapters have been devoted to a consideration of thermodynamics but rigour in treatment has been deliberately diluted in an attempt to aid comprehension.

The bulk of the book contains a discussion on binary, ternary and quaternary systems. The binary systems are treated fully, ternaries selectively and quaternaries highly selectively. The mastery of ternary and multi-component systems depends on an ability to think three-dimensionally. The serious student will find it helpful to re-draw many of the space diagrams for himself and to construct his own series of sections through them. The use of simple wire models and the moulding of surfaces and spaces from plasticene are effective teaching aids.

At appropriate points throughout the book reference is made to actual alloy systems. Not only does this relieve the tedium of an abstract treatment in terms of components A, B, etc., but it is hoped that it may induce the reader to consult the original papers. No modern discussion of multi-component systems would be complete without a presentation of recent Russian work on the application of topo-analytical methods to the study of phase diagrams. Chapter 14 introduces this approach, but those who wish to proceed further in this fascinating field could well consult the textbook by L. S. PALATNIK and A. I. LANDAU recommended on page 281.

Many individuals and organisations have willingly helped in the preparation of this book. I would particularly like to acknowledge the constructive criticisms of Professor H. K. M. LLOYD and Mr. R. D. GARWOOD, of the University College of South Wales and Monmouthshire, Cardiff, who read the original manuscript and contributed much useful comment. I would also wish to thank Elsevier Publishing Company for undertaking to render the original sketches into more elegant illustrations.

Harpenden,   A. PRINCE
Herts., England

Acknowledgements

Grateful acknowledgement is made to the following for permission to reproduce diagrams and textual matter:

I. E. COTTINGTON, A.R.P.S., for Fig. 47b, 47c, and 52.

Dover Publications Inc., for extracts from *The Scientific Papers of J. Willard Gibbs*, Vol. 1, *Thermodynamics*, 1961. (Re-publication of 1st edn. by Longmans, Green & Co., 1906.)

Dr. Riederer-Verlag GmbH for Fig. 202 from W. KÖSTER AND K. HAUG, *Z. Metallk.*, **48** (1957) 327; Fig. 205 from W. KÖSTER AND W. UHLRICH, *ibid.*, **49** (1958) 361; Fig. 207 from R. DOBBENER, *ibid.*, **48** (1957) 413; and Fig. 217 from B. PREDEL, *ibid.*, **52** (1961) 507.

Edward Arnold (Publishers) Ltd., for Figs. 44, 69, and 73 from A. H. COTTRELL, *Theoretical Structural Metallurgy*, 1948.

Elsevier Publishing Co., for Fig. 67b.

McGraw-Hill Publishing Co. Ltd., for Fig. 122 from F. N. RHINES, *Phase Diagrams in Metallurgy*, 1956.

Metallurgical Services of Betchworth, Surrey, for Fig. 28b, 91b and 180.

Oliver and Boyd Ltd., for Fig. 25 from M. H. RAND AND O. KUBASCHEWSKI, *The Thermochemical Properties of Uranium Compounds*, 1963.

Reinhold Publishing Co. Ltd., for an extract from G. MASING, *Ternary Systems*, translated by B. A. ROGERS, 1944.

The American Chemical Society for Fig. 56 from C. D. THURMOND AND J. D. STRUTHERS, *J. Phys. Chem.*, **57** (1953) 832.

The American Society for Metals for Fig. 89a and 92 from *Metals Handbook*, 1948 edn.

The British Iron and Steel Research Association for Fig. 91a.

The British Non-Ferrous Metals Research Association for Fig. 80a.

The Copper Development Association for Fig. 89b.

The Institute of Metals for Fig. 71 from G. V. RAYNOR, *Annotated Equilibrium Diagram Series No. 3;* Fig. 95 from R. P. JEWETT AND D. J. MACK, *J. Inst. Metals*, **92** (1963–64) 59; Fig. 96 from W. HUME-ROTHERY AND G. V. RAYNOR, *The Structure of Metals and Alloys*, 4th. edn., 1962; Fig. 97a from G. V. RAYNOR, *Annotated Equilibrium Diagram Series No. 2.*

The Tin Research Institute for Fig. 47a.

Contents

CHAPTER 1

Introductory Thermodynamics

1.1. EQUILIBRIUM

The concept of stable equilibrium is of fundamental importance. We can define various types of equilibrium. Mechanical equilibrium is a condition achieved when all particles are at rest and the total potential energy of the system is a minimum. An apt analogy is that of the match-box. When on its face it is in stable mechanical equilibrium (Fig. 1a). When balanced on an edge it is in unstable equilibrium. Note that when placed on the narrow face (Fig. 1c), the matchbox is not in stable mechanical equilibrium since it can reduce its potential energy by reverting to position a. Position c is referred to as metastable mechanical equilibrium.

Thermal equilibrium is the condition resulting from the absence of temperature gradients in the system. Chemical equilibrium is obtained when no further reaction occurs between the reacting substances, $i.e.$ the forward and reverse rates of reaction are equal. Finally, thermo-dynamic equilibrium occurs when the system is under mechanical, thermal and chemical equilibrium. The properties of the system—pressure, temperature, volume and concentrations—do not change with time.

Thermodynamically a system is said to be in equilibrium when its free energy is a minimum. The free energy under constant pressure conditions, as is usual in alloy equilibria, is defined as

$$G = H - TS \tag{1}$$

where G is the Gibbs free energy, H is the heat content or enthalpy, and S is the entropy. Under conditions of constant volume the Helmholtz free energy is used

$$F = E - TS \tag{2}$$

where E is the internal energy.

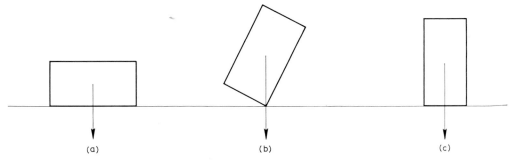

Fig. 1. Mechanical equilibrium. (a) Stable; (b) unstable; (c) metastable.

1.2. INTERNAL ENERGY

The internal energy E of a system is the sum of the kinetic and potential energies of all parts of the system. If the system were completely isolated from its surroundings (a closed system) the internal energy of the system would remain constant. If, on the other hand, we allow the system to react with its surroundings its internal energy will change. Assume that we change the state of the system (its thermal and mechanical condition) from state A to state B and that this is accompanied by an increase in internal energy dE. If this change of state is brought about by the extraction of heat δQ from the surroundings and the simultaneous performance of work δW by the system on the surroundings, the First Law of Thermodynamics (the Law of Conservation of Energy) states that the increase in internal energy is dE, where

$$dE = \delta Q - \delta W. \tag{3}$$

Note that dE is an exact differential since E is a function of the state of the system. δQ and δW are inexact differentials since their values depend on the path by which the state of the system changes from A to B, whereas the internal energy is a function of the initial and final states only.

As an example of the First Law, consider a gas at a pressure P obtaining heat from its

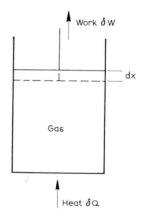

Fig. 2. The First Law of Thermodynamics applied to the expansion of a gas.

surroundings. The gas expands and does work in moving a piston of area A upwards by an amount dx (Fig. 2). The work done $\delta W = P \cdot A \cdot dx = P \cdot dV$. The First Law can therefore be re-written in the form

$$dE = \delta Q - P \cdot dV. \tag{4}$$

1.3. HEAT CONTENT OR ENTHALPY

The heat content or enthalpy H of a system is related to its internal energy by the equation
$$H = E + PV. \tag{5}$$

The enthalpy, like the internal energy, of a system is a function of the state of the system.

The heat capacity of a substance is the amount of heat required to raise its temperature by one degree

i.e.,
$$C = \frac{\delta Q}{dT}.$$ (6)

Dividing eqn. (4) by dT:

$$\frac{dE}{dT} = \frac{\delta Q}{dT} - P\frac{dV}{dT}.$$ (7)

At constant volume V

$$C_V = \left(\frac{\delta Q}{dT}\right)_V = \left(\frac{dE}{dT}\right)_V.$$ (8)

To obtain the heat capacity at constant pressure, C_P, we differentiate eqn. (5) to give

$$dH = dE + P\,dV + V\,dP.$$ (9)

Substituting eqn. (4) for dE,

$$dH = \delta Q + V\,dP.$$ (10)

Dividing by dT,

$$\frac{dH}{dT} = \frac{\delta Q}{dT} + V\frac{dP}{dT}.$$ (11)

At constant pressure,

$$C_P = \left(\frac{\delta Q}{dT}\right)_P = \left(\frac{dH}{dT}\right)_P.$$ (12)

From eqn. (10) it is seen that for constant pressure $dH = \delta Q$, whereas for constant volume eqn. (4) yields the relation $dE = \delta Q$.

As stated, in alloy work one usually considers systems at constant pressure so that enthalpy changes are more important than energy changes.

1.4. ENTROPY

Tin can exist in two allotropic forms. White tin with a tetragonal structure is stable above 13 °C and the cubic gray tin is stable below 13 °C. The heat content of substances is referred to a standard state which, for solids, is taken to be one atmosphere pressure and a temperature of 25 °C. The heat content in the standard state is taken to be zero. Thus white tin, the stable allotrope at 25 °C, has a heat content equal to zero. Gray tin has a heat content at 25 °C of −0.5 kcal/mole. A negative value corresponds to an evolution of heat in the transformation from white to gray tin at 25 °C, *i.e.* white tin $\overset{25°C}{\rightleftharpoons}$ gray tin involves the evolution of 500 cal/ mole. This implies that gray tin is more stable than white tin at 25 °C, in line with the assumption that a system generally becomes more stable as its heat content decreases. As we know, however, white tin is the stable allotrope at 25 °C. The sign of the heat content change is not a sufficient criterion for determining the course of a reaction. The criterion which is more appropriate is the sign of the entropy change for the reaction.

Entropy is an extensive quantity. That is to say, it is halved when the system is divided into two. Similarly, the internal energy and volume of a system are extensive quantities but the pressure is an intensive quantity.

Entropy can be considered from two distinct points of view—the classical thermodynamic one in which entropy is a thermodynamic variable of the system, and the atomistic or statistical mechanical approach in which entropy is a measure of the number of ways of arranging the atoms or molecules composing the system. The thermodynamic approach is rather abstract and gives little grasp of the meaning of entropy. The atomistic approach leads to identical conclusions and it has the great advantage of being comprehensible in terms the metallurgist can appreciate.

1.4.1. *Atomistic concept*

To define any system on the atomic scale would require a statement of the position and velocity of every atom or molecule in the system. Statistical mechanics is concerned with the application of statistical methods to measure the average macroscopic properties of the assembly of atoms in the system. Consider the case where we have four coloured balls—blue, green, red and yellow—and two boxes. The number of different ways of distributing the four balls in the two boxes is 2^4, *i.e.*,

Box 1	bgry	gry	bry	bgy	bgr	ry	gy	gr	by	br	bg	y	r	g	b	
Box 2		b	g	r	y	bg	br	by	gr	gy	ry	bgr	bgy	bry	gry	bgry

There are sixteen different ways, or complexions, for arranging the balls and if the balls are distributed in a random way each complexion has the same probability of occurring. In this way all four balls have as much chance of finding themselves in box 2 as being divided equally between boxes 1 and 2 such that b and g are in box 1 and r and y in box 2 for example.

However, if we examine the sixteen ways the balls can distribute themselves between the two boxes we see that, if the balls are distributed in the boxes in a random manner so that any of the sixteen complexions are equally probable, the probability of achieving a uniform distribution of the balls between the boxes (two balls in each box) is greater than the probability of any other type of distribution.

The probability of finding 4 balls in box 1   1 : 16
3 balls in box 1 and 1 ball in box 2   4 : 16
2 balls in each box   6 : 16
1 ball in box 1 and 3 balls in box 2   4 : 16
4 balls in box 2   1 : 16

Had we taken six balls then the total number of ways of distributing them would be $2^6 = 64$.

The probability of finding 6 balls in box 1   1 : 64
5 balls in box 1 and 1 ball in box 2   6 : 64
4 balls in box 1 and 2 balls in box 2   15 : 64
3 balls in each box   20 : 64
2 balls in box 1 and 4 balls in box 2   15 : 64
1 ball in box 1 and 5 balls in box 2   6 : 64
6 balls in box 2   1 : 64

With ten balls the probability of finding all 10 balls in box 1 would be 1 : 1024, and the probability of finding 5 balls in each box 252 : 1024. With twenty balls the probability of finding all 20 balls in one box would be 1 : 1,048,576, and the probability of finding 10 balls in each box 184,756 : 1,048,576. These examples illustrate quite effectively that the chance of obtaining a more random distribution of the balls increases greatly as the number of balls is increased from four to a mere twenty. In fact with twenty balls the chance of finding all 20 balls in one particular box is almost negligible—1 in a million chance.

A similar argument may be applied to the mixing of two gases initially separated by a partition. If the partition is removed the gases mix rapidly but the reverse process of unmixing of two gases is never observed. This is so because, although the two gases have no preference for any particular distribution of their constituent molecules (*i.e.* each distribution has the same probability) the homogeneous mixture of the two gases (corresponding to a completely random distribution of their molecules) corresponds to such a hugely superior number of complexions than the more orderly distribution of molecules that the latter never exists.

The term entropy can be introduced to measure the probability of any given distribution of the atoms in a system; the most probable distribution has the highest entropy. The statistical definition of entropy is given by Boltzmann's relation:

$$S = k \ln p \tag{13}$$

where k is Boltzmann's constant (the gas constant per molecule, or R/N_0, R being the gas constant and N_0 Avogadro's number), p is the probability of the given distribution and S is the entropy. As defined, entropy has the dimension of energy/temperature since $k = 1.3804 \times 10^{-16}$ ergs/degree. Energy itself has the dimension ml^2t^{-2} and so entropy has the dimension $ml^2t^{-2}T^{-1}$. The thermodynamic definition of entropy (p. 7) leads to the same dimension. We have used p so far in the sense of the number of ways of distributing the atoms in the system (or balls in the boxes). We have also shown that as the number of atoms or balls increases the most probable distribution of atoms centres increasingly on a completely random or disordered arrangement. This uniform distribution can be regarded as the most random state possessing the highest entropy. Entropy is therefore a measure of the randomness or disorder in a system.

As one might expect, the formation of a solid solution by mixing atoms of A with atoms of B introduces randomness or disorder into the system with a consequent increase in the entropy of the system. This entropy increase consequent upon the formation of a solid solution is called the entropy of mixing. It arises because the crystal composed of pure A atoms (or pure B atoms) contains only atoms of one kind, each indistinguishable from its neighbours. There is therefore only one way of distributing the A atoms (or B atoms) on the lattice sites in the crystal of pure A (or B). In the A–B solid solution there will be many ways of distributing the A and B atoms in a completely random manner. Take as an example the distribution of four atoms on four lattice sites. If all the atoms are A atoms and each atom is considered to be indistinguishable from its neighbours, there is only one way of distributing the four A atoms on the four lattice sites, *viz.*

A	A
A	A

If there are three A atoms and one B atom they can be distributed on the four lattice sites in four ways, *viz.*

Two A and two B atoms can be distributed in six different ways, *viz.*

One A and three B atoms can be distributed in four ways, in a manner similar to that illustrated for one B and three A atoms. If all four atoms are B atoms there is only one way of distributing them, as shown for the case of four A atoms on four lattice sites. In general, if there are N lattice sites and n atoms of A and $N-n$ atoms of B, there are $N!/[n!(N-n)!]$ ways of distributing the atoms on the lattice sites.

Boltzmann's relation can be used to determine the additional entropy, ΔS_m, on forming the A–B solid solution:

$$nA+(N-n)B = [nA, (N-n)B]_{solution}$$

i.e. n atoms of A added to $(N-n)$ atoms of B to form a solid solution containing n atoms of A and $(N-n)$ atoms of B.

$$\Delta S_m = S_{A-B}-S_A-S_B$$
$$\Delta S_m = k(\ln p_{A-B} - \ln p_A - \ln p_B)$$

or

$$\Delta S_m = k \ln p_{A-B}$$

since

$$\ln p_A = \ln p_B = \ln 1 = 0$$

$$\Delta S_m = k \ln \frac{N!}{n!(N-n)!}. \tag{14}$$

Using the approximation (Stirling's Theorem*),

$$\ln N! = N \ln N - N \tag{15}$$
$$\Delta S_m = k[N \ln N-(N-n) \ln n+n-(N-n) \ln (N-n)+(N-n)]$$
$$= k[N \ln N-n \ln n-(N-n) \ln (N-n)]. \tag{16}$$

Denoting the atomic fraction of A in the solid solution as X_A and the atomic fraction of B as $(1-X_A) = X_B$, then

$$X_A = \frac{n}{N} \quad \text{and} \quad (1-X_A) = X_B = \frac{(N-n)}{N}.$$

In terms of concentration

$$\Delta S_m = -Nk[X_A \ln X_A+(1-X_A) \ln (1-X_A)]. \tag{17}$$

---

* $\log N! = \log N + \log (N-1) + \log (N-2) +\ .\ .\ .+ \log 1 = \Sigma_1^N \log N.$
For large values of N a continuous function can be assumed. Thus $\Sigma_1^N \log N = \int_0^N \log N \cdot dN = N \log N-N.$

Equation (17) is the expression for the entropy of mixing in a solid solution. If we consider a gram-atom, then N is equal to Avogadro's number (6.023×10^{23}) and $Nk = R = 1.986$ cal/degree K, and

$$\Delta S_m = -R[X_A \ln X_A + (1 - X_A) \ln (1 - X_A)]. \tag{18}$$

As X_A and $(1 - X_A)$ are fractions their logarithms are negative, and hence ΔS_m is positive. In this idealised case the entropy of mixing when plotted against concentration gives a curve which is symmetrical about $X_A = 0.5$ (Fig. 3).

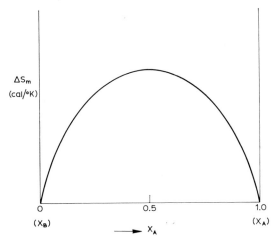

Fig. 3. Variation of the entropy of mixing of an ideal solution with composition.

The maximum value of ΔS_m for $X_A = 0.5$ can be calculated using eqn. (18). It is 1.377 cal/degree K. The symmetrical curve for $\Delta S_m - X_A$ would be expected from the previous discussion of the distribution of A atoms and B atoms on lattice sites. In the example considered, the maximum number of ways for distributing A and B atoms was associated with a uniform sharing of sites between A and B atoms. This state has the highest probability and therefore the largest entropy.

1.4.2. Thermodynamic definition

The thermodynamic definition of entropy arose from a consideration of the conditions under which heat can be converted into work as, for example, in the steam engine. When considering the First Law of Thermodynamics it was stated that the amount of heat necessary to change a system from an equilibrium state A to another equilibrium state B was dependent on the way in which the equilibrium is changed, $i.e.$ on the path followed from A to B. This amount of heat,

$$Q_B - Q_A = \int_A^B \delta Q.$$

A quantity which is uniquely defined by the state of the system is the integral

$$\int_A^B \frac{\delta Q}{T},$$

where T is the absolute temperature and the system is assumed to undergo a reversible transition from state A to state B. From a consideration of Carnot cycles it can be shown that in a reversible process or cycle

$$\int_A^B \frac{\delta Q}{T} = 0.$$

In an irreversible process or cycle

$$\int_A^B \frac{\delta Q}{T} < 0.$$

A new thermodynamic function, S, is defined such that $dS = \delta Q/T$ for a reversible process. The function S is known as the entropy. For a reversible change of state,

$$S_B - S_A = \Delta S = \int_A^B \frac{\delta Q}{T}.$$

To give a fixed reference point the entropy of a pure substance at 0 °K is taken to be zero. The entropy at T °K is then

$$S = \int_0^T \frac{\delta Q}{T}.$$

The value of S at T °K will be independent of the path by which the system has reached this state.

In a system which can exchange heat with its surroundings the entropy change of the system must be considered. Take such a reversible process as the transformation of gray tin to white tin. From eqn. (5)

$$H = E + PV.$$

Differentiating,

$$dH = dE + P\,dV + V\,dP.$$

From eqn. (4)

$$dE = \delta Q - P\,dV.$$

At constant pressure

$$\delta Q = dH.$$

Therefore

$$dS = \frac{\delta Q}{T} = \frac{dH}{T}$$

$$\int_{gray}^{white} dS = \frac{1}{T} \int_{gray}^{white} dH.$$

Integrating at constant temperature,

$$\Delta S = \frac{\Delta H_t}{T}$$

where ΔH_t is the enthalpy change for the transformation of gray tin to white tin. Since $\Delta H_t = 500$ cal/mole and the transformation temperature is 13 °C (286 °K), $\Delta S = \frac{500}{286} = 1.75$ cal/mole degree.

The entropy increase of the tin is 1.75 cal/mole degree. The surroundings must lose an equivalent amount of entropy. In the transformation 500 calories entered the system from the surroundings and, since the transformation is isothermal,

$$\Delta S_{\text{surroundings}} = -\frac{\Delta H_t}{T} = -1.75 \text{ cal/mole degree.}$$

The total entropy change in the reversible process is zero

$$\Delta S + \Delta S_{\text{surr}} = 0.$$

In an irreversible process or cycle $\int \delta Q/T < 0$. Consider an irreversible cycle whereby a system is transferred from state A to state B in an irreversible manner and from state B back to state A in a reversible manner. The cycle as a whole is irreversible and we can write

$$\underbrace{\int \frac{\delta Q}{T}}_{\substack{\text{completely} \\ \text{irreversible} \\ \text{cycle}}} = \underbrace{\int_A^B \frac{\delta Q}{T}}_{\substack{\text{irreversible} \\ A \rightarrow B}} + \underbrace{\int_B^A \frac{\delta Q}{T}}_{\substack{\text{reversible} \\ B \rightarrow A}} < 0.$$

But

$$\underbrace{\int_B^A \frac{\delta Q}{T}}_{\substack{\text{reversible} \\ B \rightarrow A}} = S_A - S_B.$$

Therefore

$$\underbrace{\int_A^B \frac{\delta Q}{T}}_{\substack{\text{irreversible} \\ A \rightarrow B}} < S_B - S_A. \tag{19}$$

For an irreversible change in state $dS > \delta Q/T$ in contrast to the reversible change with $dS = \delta Q/T$. To illustrate the fact that the change in entropy in an irreversible process is greater than $\int \delta Q/T$, one can consider the melting of ice. If ice is melted reversibly it receives latent heat of fusion from the surroundings, which are substantially at 0 °C. To melt ice irreversibly one could surround the ice by water at 20 °C. The same latent heat would be involved in melting the ice but, since T has increased from 0 °C to 20 °C, the quantity Q/T will be smaller than the entropy change.

Instead of considering the system and its surroundings separately, we can consider them jointly as an isolated system. In an isolated system all the bodies concerned in the exchange of energy are considered as part of the system. The system is thought to be contained in a rigid adiabatic envelope. For an irreversible change of state under adiabatic conditions eqn. (19) now becomes

$$S_B - S_A > 0 \quad \text{or} \quad S_B > S_A$$

since $\delta Q/T$ is zero as no heat can enter or leave an isolated system. For a reversible process under adiabatic conditions $dS = 0$.

To summarise:

	Non-isolated systems	Isolated systems	
Reversible	$dS = \dfrac{\delta Q}{T}$	$dS = \dfrac{\delta Q}{T} = 0$	where dS is the entropy change of the *system* only.
Irreversible	$dS > \dfrac{\delta Q}{T}$	$dS > \dfrac{\delta Q}{T} > 0$	

The consideration of isolated systems leads to an important conclusion. The entropy of such systems either remains constant ($dS = 0$) or increases ($dS > 0$). The entropy remains constant when the system is in an equilibrium state; if it is not, the entropy increases until equilibrium is reached. This is an expression of the Second Law of Thermodynamics—the entropy in an isolated system tends to a maximum. This concept is a valuable one for distinguishing the direction in which reactions can take place. A reaction in equilibrium has $dS = 0$, a spontaneously occurring reaction is equated with $dS > 0$, and an impossible reaction with $dS < 0$.

The principle of the increase in entropy of a system refers only to isolated systems. As we have noted, for a system which can exchange heat with its surroundings there can be either an increase or decrease in entropy of the system but this is counterbalanced by a decrease or an increase in the entropy of the surroundings. The transition of gray tin to white tin was used as an example of the system gaining entropy and the surroundings losing an equivalent amount of entropy. The reverse case can be simply illustrated by the cooling of a beaker of water from 100 °C to room temperature. Its entropy decreases but the surroundings gain the entropy lost by the system. From the statistical approach to entropy, the statement that the entropy of an isolated system tends to a maximum is equivalent to saying that the system tends to a state which is the most probable one, *i.e.* there is a maximum number of ways of realising this state. For formal proof of the equivalence of the statistical and thermodynamic definitions of entropy the reader is referred to standard thermodynamic texts.

1.5. FREE ENERGY

An isolated system is in equilibrium when its entropy is a maximum. In metallurgical work we are mainly concerned with examining the reactions within a system that is not isolated from its surroundings as, for example, in the thermal analysis of alloys. It is a distinct drawback to the use of the entropy concept to have to consider not only the entropy changes of the system but also that of its surroundings. What is required is some function of a system which we can use to define the conditions of equilibrium of the system under all conditions in the way that entropy can be used to define equilibrium for isolated systems. The function used is called the free energy.

Consider a system which can exchange energy with its surroundings. For such a system we have noted that

$$dS + dS_{surr} \geq 0. \tag{20}$$

where dS refers to the system. Suppose the system absorbs heat δQ from its surroundings at temperature T. Then

$$dS_{surr} = -\frac{\delta Q}{T}. \tag{21}$$

Under isobaric conditions eqn. (10) indicates that

$$\delta Q = dH$$

Therefore

$$dS_{surr} = -\frac{dH}{T}. \tag{22}$$

From eqns. (20) and (22)

$$dS - \frac{dH}{T} \geq 0$$

or

$$dH - T \, dS \leq 0. \tag{23}$$

Differentiating,

$$d(H - TS) = dH - T \, dS - S \, dT$$

$$= dH - T \, dS \text{ for isothermal conditions.}$$

Therefore

$$d(H - TS)_{T, P} \leq 0. \tag{24}$$

Under isothermal and isobaric conditions the expression $(H - TS)$ defines the conditions for equilibrium in the system. For a reversible reaction $H - TS = 0$. For an irreversible reaction $H - TS < 0$. The expression $(H - TS)$ is called the Gibbs free energy, the free enthalpy, the thermodynamic potential or the free energy at constant pressure. It is denoted by the symbol G. The tendency for the entropy to a maximum can therefore be replaced by the more useful tendency of the Gibbs free energy to a minimum. In a reversible reaction, such as an allotropic transformation, the change of Gibbs free energy for the transformation is zero. Both allotropes co-exist in equilibrium. In an irreversible reaction the change in the Gibbs free energy is negative; such a process is accompanied by a decrease in free energy.

If, instead of isothermal isobaric conditions, we chose isothermal and constant volume conditions, then

$$dS_{surr} = -\frac{\delta Q}{T} = \frac{dE}{T}$$

$$dS - \frac{dE}{T} \geq 0$$

$$dE - T \, dS \leq 0$$

or

$$d(E - TS)_{T, V} \leq 0. \tag{25}$$

The expression $(E - TS)$ is called the Helmholtz free energy or the free energy at constant volume. It is denoted by the symbol F.

Under equilibrium conditions,

(1) at a given temperature and pressure the Gibbs free energy $G = H - TS$ is at a minimum, and the change in free energy $dG_{TP} = d(H - TS)_{TP} = 0$;

(2) at a given temperature and volume the Helmholtz free energy $F = E - TS$ is at a minimum, and the change in free energy $dF_{TV} = d(E - TS)_{TV} = 0$.

For irreversible processes,

$$dG_{TP} = d(H - TS)_{TP} < 0$$
$$dF_{TV} = d(E - TS)_{TV} < 0.$$

We can divide these quantities into two separate functions:

$$dH_{TP} < 0 \quad \text{or} \quad dE_{TV} < 0$$

and

$$dS_{TP} > 0 \quad \text{or} \quad dS_{TV} > 0.$$

The energy and entropy functions can be thought of as mathematical formulations of two opposing tendencies in the system. On the one hand a striving for a minimum energy and on the other hand a striving for maximum entropy. At the absolute zero the entropy term is zero and the stable state of the system is that with the lowest energy (H or E). This state generally corresponds with a degree of order in the system. A pure metal at 0 °K is a solid rather than a liquid—ordered rather than disordered. As the temperature is increased, the order in the solid decreases due to increasingly violent vibration of the atoms. The entropy function becomes predominant with a rise in temperature since this favours the disordered state. The temperature at which the entropy begins to predominate will depend on the magnitude of the attractive forces between the atoms. At the melting point the two opposing tendencies cancel each other out since $\Delta H = T\Delta S$ or $\Delta E = T\Delta S$.

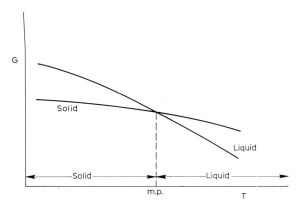

Fig. 4. Variation of free energy with temperature for a pure metal.

We can plot the free energy changes schematically. Figure 4 illustrates such a plot of G–T for a pure metal. At the melting point the free energy of liquid and solid are equal; below the melting point the solid has the lower free energy and is therefore the stable phase; above the melting point the reverse is the case and the liquid is the stable phase. It is not immediately obvious why the free energy of a phase decreases with increasing temperature. If the phase increases in temperature from T to $T + dT$ at constant pressure the heat absorbed by the system is $dQ = C_p \, dT$, where C_p is the heat capacity. Using eqn. (12) the heat content, dH, of the system will also increase by $C_p \, dT$. The entropy increase is given by $dS = dQ/T = (C_p/T) \, dT$. The change in free energy of the system, dG, with increase in temperature dT will be

$$dG = dH - T \, dS - S \, dT = C_p \, dT - T \frac{C_p}{T} \, dT - S \, dT = -S \, dT$$

where S is the entropy of the system at temperature T. The free energy of the system at temperature T is

$$G = G_0 - \int_0^T S \, dT$$

where G_0 is the free energy at the absolute zero. From eqn. (1) $G_0 = H_0$.

Therefore

$$G = H_0 - \int_0^T S \, dT.$$

But

$$S = S_0 + \int_0^T dS = S_0 + \int_0^T \frac{C_p}{T} \, dT = \int_0^T \frac{C_p}{T} \, dT$$

since S_0, the entropy at $0°$ K is zero.

Therefore

$$G = H_0 - \int_0^T \left(\int_0^T \frac{C_p}{T} \, dT \right) dT. \tag{26}$$

As H_0 is a constant, it follows from eqn. (26) that the free energy of a phase decreases with increase of temperature. A further consequence of this relation is that the phase stable at high temperatures is that which has the larger specific heat.

Certain thermodynamic relations will be useful in other chapters. Starting with the relation

$$G = H - TS \tag{27}$$

by differentiation we obtain

$$dG = dH - T \, dS - S \, dT$$

From eqn. (9)

$$dG = dE + P \, dV + V \, dP - T \, dS - S \, dT.$$

From eqn. (4)

$$dG = \delta Q + V \, dP - T \, dS - S \, dT.$$

Since $dS = \dfrac{\delta Q}{T}$

$$dG = V \, dP - S \, dT.$$

Dividing by dP, at constant temperature

$$\left(\frac{\partial G}{\partial P} \right)_T = V. \tag{28}$$

One can derive other relations such as

$$\left(\frac{\partial G}{\partial T} \right)_P = -S \tag{29}$$

and

$$\left[\frac{\partial \left(\dfrac{G}{T} \right)}{\partial \left(\dfrac{1}{T} \right)} \right]_P = H \tag{30}$$

CHAPTER 2

Thermodynamics of Solutions

2.1. SOLUTIONS

2.1.1. *Partial molar quantities*

Alloys can be considered from an elementary point of view as solution of one metal in another. If we mix two molten metals together the total volume of the solution formed is generally not equal to the sum of the individual volumes before mixing. A contraction in volume will occur if the attractive forces between the two metal atoms in the solution is greater than the attractive force in either pure metal. There is no way of assigning a proportion of such a volume contraction to each component metal in the solution. This type of problem is avoided by introducing the concept of partial molar quantities for all extensive thermodynamic functions (volume, internal energy, enthalpy, entropy and free energy).

Taking an arbitrary amount of solution of volume V' containing n_A moles* of metal A and n_B moles of metal B, the molar volume of the solution, V, is given as

$$V = \frac{V'}{n_A + n_B}. \tag{31}$$

If we add to this solution of volume V' a very small number of moles, dn_A, of metal A at constant pressure and temperature then the volume V' increases by dV'. Its rate of increase with addition of metal A, $(\partial V'/\partial n_A)_{P,T,n_B}$, is known as the partial molar volume of A. Partial molar quantities are distinguished by a bar sign, *viz.* $\overline{V}_A$.

$$\overline{V}_A = \left(\frac{\partial V'}{\partial n_A}\right)_{P,T,n_B} \tag{32}$$

As V' is a function of n_A and n_B

$$dV' = \left(\frac{\partial V'}{\partial n_A}\right)_{P,T,n_B} dn_A + \left(\frac{\partial V'}{\partial n_B}\right)_{P,T,n_A} \cdot dn_B$$

or

$$dV' = \overline{V}_A \cdot dn_A + \overline{V}_B \cdot dn_B. \tag{33}$$

It is assumed that the addition of dn_A moles of A to the solution causes no change in composition of the solution, *i.e.*, dn_A is regarded as an infinitesimal addition. Alternatively, if dn_A is not infinitesimally small we assume no change in composition of the solution if its volume V' is infinitely large. In this case if n_A moles of A are added to the solution the total volume will increase by $n_A \overline{V}_A$ from eqn. (32). If now n_B moles of B are added the volume will increase further

---

* One mole is one gram molecular weight or, with metals, one gram atom.

by $n_B \overline{V}_B$. The total change in volume is independent of the order in which the metals are added and will be equal to V' since all we have done is to produce an additional quantity of the same solution containing n_A moles of A and n_B moles of B.

Hence

$$V' = \overline{V}_A n_A + \overline{V}_B n_B. \tag{34}$$

If eqn. (34) is differentiated we obtain

$$dV' = \overline{V}_A \, dn_A + n_A \, d\overline{V}_A + \overline{V}_B \, dn_B + n_B \, d\overline{V}_B. \tag{35}$$

Recalling eqn. (33), we find

$$n_A \, d\overline{V}_A + n_B \, d\overline{V}_B = 0. \tag{36}$$

An arbitrary volume, V', of solution has been considered so far. It is more convenient to use one mole of solution. In alloys this is equivalent to using a gram-atom of solution. To obtain the molar volume, V, eqn. (31) indicates that we divide by the number of moles in the solution. Equations (33), (34) and (36) can be re-written in the form:

$$\frac{dV'}{d(n_A + n_B)} = \overline{V}_A \cdot \frac{dn_A}{d(n_A + n_B)} + \overline{V}_B \cdot \frac{dn_B}{d(n_A + n_B)} \tag{37}$$

$$\frac{V'}{n_A + n_B} = \overline{V}_A \cdot \frac{n_A}{n_A + n_B} + \overline{V}_B \cdot \frac{n_B}{n_A + n_B} \tag{38}$$

$$\frac{n_A}{n_A + n_B} \cdot d\overline{V}_A + \frac{n_B}{n_A + n_B} \cdot d\overline{V}_B = 0. \tag{39}$$

But the mole fraction, or atomic fraction for alloys, is defined as follows

$$X_A = \frac{n_A}{n_A + n_B}; \quad X_B = \frac{n_B}{n_A + n_B} = 1 - X_A.$$

Therefore, eqns. (37), (38) and (39) reduce to

$$dV = \overline{V}_A \, dX_A + \overline{V}_B \, dX_B \tag{40}$$

$$V = \overline{V}_A X_A + \overline{V}_B X_B \tag{41}$$

$$X_A \, d\overline{V}_A + X_B \, d\overline{V}_B = 0. \tag{42}$$

Similar sets of equations can be derived for the other extensive thermodynamic functions—E, H, S and G or F. Partial molar quantities can easily be derived from molar quantities. If the molar volume, or volume per gram-atom, has been determined as a function of composition (Fig. 5a), the partial molar volume of A and B in the solution, $\overline{V}_A$ and $\overline{V}_B$, can be derived from the molar volume by drawing a tangent to the curve at the composition under consideration. In Fig. 5a, at composition X the solution has a molar volume V. The tangent to the curve at this molar volume intersects the $X_A = 1$ and $X_B = 1$ ordinates at the values of the partial molar volumes of A and B respectively. This statement can be proved by taking eqn. (40) and multiplying through by X_A/dX_B to give

$$X_A \frac{dV}{dX_B} = -X_A \overline{V}_A + X_A \overline{V}_B \tag{43}$$

noting that $dX_A = -dX_B$ since $X_A + X_B = 1$.

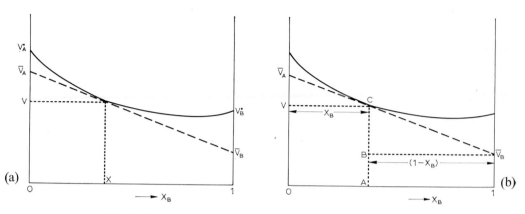

Fig. 5. Derivation of partial molar volumes from molar volumes.

Adding eqns. (41) and (43),

$$V + X_A \frac{dV}{dX_B} = \overline{V}_B(X_A + X_B) = \overline{V}_B.$$

Therefore

$$\overline{V}_B = V + (1 - X_B) \frac{dV}{dX_B}$$

Similarly,

$$\overline{V}_A = V - X_B \frac{dV}{dX_B}$$

From Fig. 5b,

$$\overline{V}_B = \overline{AC} + (1 - X_B)\left(-\frac{\overline{BC}}{1 - X_B}\right)$$
$$= \overline{AC} - \overline{BC} = \overline{AB}.$$

i.e., the ordinate of the tangent at $X_B = 1$ is $\overline{V}_B$. Similarly for $\overline{V}_A$.

The molar volumes of the pure metals A and B are represented in Fig. 5a by the terms $V_A^\bullet$ and $V_B^\bullet$ respectively. It will be noted that the partial molar volumes of A and B in the solution vary with the composition of the solution since different tangents intersect the ordinates at different positions.

Partial molar quantities can be inter-related for solutions of a fixed composition. Naturally the solution has to be of a fixed composition since we have seen that the partial molar quantities of the components in the solution vary with the composition of the solution.

2.1.2. Relative molar quantities

The free energy of an arbitrary quantity of solution can be written $G' = H' - TS'$.

By differentiation

$$\left(\frac{\partial G'}{\partial n_A}\right)_{P,\,T,\,n_B} = \left(\frac{\partial H'}{\partial n_A}\right)_{P,\,T,\,n_B} - T\left(\frac{\partial S'}{\partial n_A}\right)_{P,\,T,\,n_B}.$$

From eqn. (32)

$$\overline{G}_A = \overline{H}_A - T\overline{S}_A. \tag{44}$$

The relation between the partial molar thermodynamic quantities is identical to the relations previously given for pure components. Equation (44) is of identical form to eqn. (27). Similarly,

$$\left(\frac{\partial \bar{G}_A}{\partial P}\right)_{T, n_B} = \bar{V}_A \tag{45}$$
$$(cf.\ 28)$$

$$\left(\frac{\partial \bar{G}_A}{\partial T}\right)_{P, n_B} = -\bar{S}_A \tag{46}$$
$$(cf.\ 29)$$

$$\left[\frac{\partial \left(\frac{\bar{G}_A}{T}\right)}{\partial \left(\frac{1}{T}\right)}\right]_{P, n_B} = \bar{H}_A. \tag{47}$$
$$(cf.\ 30)$$

Subtracting eqn. (28) from (45) gives:

$$\frac{\partial (\bar{G}_A - G_A^{\bullet})}{\partial P} = (V_A - V_A^{\bullet}). \tag{48}$$

Similarly,

$$\frac{\partial (\bar{G}_A - G_A^{\bullet})}{\partial T} = -(\bar{S}_A - S_A^{\bullet}). \tag{49}$$

$$\frac{\partial \left(\frac{(G_A - G_A^{\bullet})}{T}\right)}{\partial \left(\frac{1}{T}\right)} = (\bar{H}_A - H_A^{\bullet}). \tag{50}$$

The quantities $(\bar{V}_A - V_A^{\bullet})$, etc. are called relative partial molar quantities. The relative partial molar volume of A is the difference between the partial molar volume of A in solution and the molar volume of pure A.

2.1.3. *Ideal solutions*

In discussing solutions we have derived the concept of partial molar quantities from a consideration of the difference between a solution and its components in the pure state (standard state). This was done by consideration of the attractive forces between two metal atoms in solution compared with the attractive forces between the atoms in the pure metals. There are obviously three possibilities:

(1) the attractive force between dissimilar atoms is equal to the attractive force between similar atoms

$$A \leftrightarrow B = \tfrac{1}{2}(A \leftrightarrow A + B \leftrightarrow B)$$

(2) the attractive force between dissimilar atoms is greater than the attractive force between similar atoms

$$A \leftrightarrow B > \tfrac{1}{2}(A \leftrightarrow A + B \leftrightarrow B)$$

(3) the attractive force between dissimilar atoms is less than the attractive force between similar atoms

$$A \leftrightarrow B < \tfrac{1}{2}(A \leftrightarrow A + B \leftrightarrow B).$$

The first case is the ideal case which would give rise to ideal solutions. The second case is the usual one; the stronger bond between dissimilar atoms reduces the effective concentrations of the two metals in solution below their actual concentrations. The reverse is true for the third case. To give expression to the effective concentrations of components in solutions the term *activity* has been introduced. It can best be understood by considering the vapour pressure of a metal when it is dissolved in another metal. The vapour pressure of the solvent is reduced from that of the vapour pressure of the pure metal, $p_A^{\bullet}$, to a value p_A. The ratio $p_A/p_A^{\bullet}$ is called the activity of metal A in solution

$$a_A = \frac{p_A}{p_A^{\bullet}}. \tag{51}$$

To return to an ideal solution. Let us examine the variation of vapour pressure of one component, A, in a binary solution. When $X_A = 1$ the vapour pressure is that of pure A, *i.e.*, $p_A^{\bullet}$. When $X_A = 0$ the vapour pressure of A in the solution is zero since there is no A present.

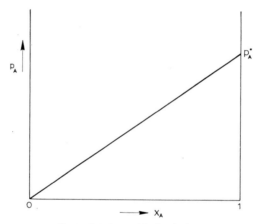

Fig. 6. Vapour pressure, p_A, of metal A in an ideal solution as a function of composition.

Between $X_A = 0$ and $X_A = 1$ the vapour pressure of A in the binary solution is a linear function of composition if the solution is ideal (Fig. 6).

Hence

$$p_A = X_A p_A^{\bullet}. \tag{52}$$

As $a_A = p_A/p_A^{\bullet}$ for an ideal solution

$$a_A = X_A. \tag{53}$$

This relation is known as Raoult's law. The activity of a solvent in an ideal solution is equal to its atomic fraction. In standard thermodynamic texts* it is shown that the activity is related to the Gibbs free energy by the equation

$$dG = RT \, d \ln a. \tag{54}$$

For a component A in a solution

$$d\bar{G}_A = RT \, d \ln a_A. \tag{55}$$

The activity of a pure component (taken as the standard reference state) is unity ($a_A^{\bullet} = 1$).

---

* For example, J. LUMSDEN, *Thermodynamics of Alloys*, Monograph No. 11, Institute of Metals, London, 1952, p. 160.

Integrating eqn. (55), we therefore obtain

$$\bar{G}_A - G_A^{\bullet} = RT \ln a_A. \tag{56}$$

For an ideal solution $a_A = X_A$ so

$$\bar{G}_A - G_A^{\bullet} = RT \ln X_A. \tag{57}$$

Substituting eqn. (57) into (50)

$$R\left[\frac{\partial \ln X_A}{\partial\left(\dfrac{1}{T}\right)}\right] = \bar{H}_A - H_A^{\bullet}. \tag{58}$$

The partial derivative on the left-hand side of eqn. (58) is zero since X_A is not a function of temperature. Therefore

$$\bar{H}_A = H_A^{\bullet}. \tag{59}$$

Equation (59) states that a component A in an ideal solution behaves as though it were in its standard state (pure) as far as enthalpy changes are concerned. Equation (41) can be re-written

$$H = \bar{H}_A X_A + \bar{H}_B X_B.$$

From eqn. (59)

$$H_{\text{ideal}} = H_A^{\bullet} X_A + H_B^{\bullet} X_B.$$

If we now consider the change in enthalpy on forming one mole of an ideal solution, $\Delta H_{m,\text{ideal}}$, (the enthalpy of mixing) it is equal to the enthalpy of the solution minus the sum of the enthalpies of the unmixed components,

i.e.,
$$\Delta H_{m,\text{ideal}} = H_{\text{ideal}} - (H_A^{\bullet} X_A + H_B^{\bullet} X_B) = 0.$$

No heat is absorbed or liberated on forming the ideal solution.

The free energy of formation (or mixing) of an ideal solution, $\Delta G_{m,\text{ideal}}$, is equal to the free energy G of the solution minus the sum of the free energies of the pure unmixed components,

$$\Delta G_{m,\text{ideal}} = G - (G_A^{\bullet} X_A + G_B^{\bullet} X_B).$$

Substituting eqn. (41), $G = \bar{G}_A X_A + \bar{G}_B X_B$,

$$\Delta G_{m,\text{ideal}} = X_A(\bar{G}_A - G_A^{\bullet}) + X_B(\bar{G}_B - G_B^{\bullet}). \tag{60}$$

Substitution of eqn. (56) yields

$$\Delta G_{m,\text{ideal}} = RT(X_A \ln a_A + X_B \ln a_B).$$

Finally, substitution of eqn. (53) gives

$$\Delta G_{m,\text{ideal}} = RT(X_A \ln X_A + X_B \ln X_B). \tag{61}$$

At constant temperature

$$\Delta G_{m,\text{ideal}} = \Delta H_{m,\text{ideal}} - T\Delta S_{m,\text{ideal}}.$$

Since $\Delta H_{m,\text{ideal}} = 0$,

$$\Delta S_{m,\text{ideal}} = -R(X_A \ln X_A + X_B \ln X_B). \tag{62}$$

This last equation for the entropy of mixing of an ideal solution is identical to that derived in a statistical manner previously (eqn. 18), recalling that $X_B = 1 - X_A$.

The volume change in the formation of an ideal solution is also zero as can be seen by combining eqns. (48) and (57). For an ideal solution the enthalpy and volume are the same as those of a purely mechanical mixture of the pure components in various amounts. The entropy of mixing of an ideal solution is a definite positive quantity, as we have already noted in Fig. 3.

2.1.4. *Actual solutions*

For an ideal solution $a_A = X_A$ is accordance with Raoult's law. There are few, if any, ideal solutions. Generally, solutions show deviations from Raoult's law. The deviation of the activity of a component in a solution is expressed by its activity coefficient, γ, defined as

$$\gamma_A = \frac{a_A}{X_A}. \tag{63}$$

An ideal solution is one with an activity coefficient equal to unity. Activity–composition curves for ideal and actual solutions are given in Fig. 7. Actual solutions (Figs. 7b and 7c) obey

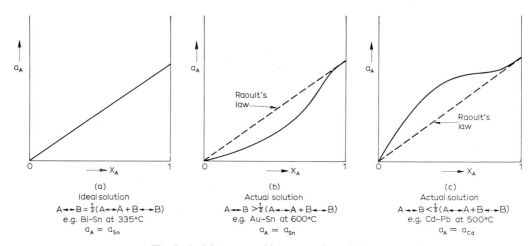

Fig. 7. Activity–composition curves for solutions.

Raoult's law at small contents of solute only. Increasing solute content (lower X_A) coincides with departure from Raoult's law. A negative deviation (Fig. 7b), with activities less than the atom fraction, is associated with a greater attractive force between A and B atoms than between similar atoms. In this way the effective concentration of A is reduced when it forms a solution with B. This type of behaviour is indicative of the formation of intermediate phases between A and B. A positive deviation (Fig. 7c) indicates a tendency to phase separation in the solid state. In the limit there would be little solubility of either component in each other, either in the liquid or solid state.

In actual solutions the solvent, A, obeys Raoult's law in dilute solutions (small contents of solute B). The solute B in dilute solutions obeys Henry's law:

$$a_B = kX_B. \tag{64}$$

Figure 8 illustrates the applicability of Raoult's and Henry's laws to actual solutions.

For an ideal solution we noted (eqn. (56)) that the partial molar free energy was $\bar{G}_A - G_A^\bullet = RT \ln a_A$. As $a_A = \gamma_A X_A$ for an actual solution, in this case

$$\bar{G}_A - G_A^\bullet = RT \ln X_A + RT \ln \gamma_A. \tag{65}$$

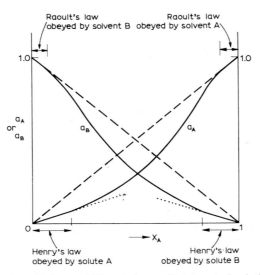

Fig. 8. Raoult's law and Henry's law applied to actual solutions.

The free energy of mixing of an actual solution, by analogy with eqn. (61), is

$$\Delta G_{m,\,\text{actual}} = RT(X_A \ln X_A + X_B \ln X_B + X_A \ln \gamma_A + X_B \ln \gamma_B)$$

or,

$$\Delta G_{m,\,\text{actual}} = RT(X_A \ln X_A + X_B \ln X_B) + RT(X_A \ln \gamma_A + X_B \ln \gamma_B).$$
$$\Delta G_{m,\,\text{ideal}} + \Delta G_{m,\,\text{excess}} \tag{66}$$

where

$$\Delta G_{m,\,\text{excess}} = RT(X_A \ln \gamma_A + X_B \ln \gamma_B). \tag{67}$$

The free energy of mixing of an actual non-ideal solution differs from the free energy of mixing of an ideal solution by the quantity on the right-hand side of eqn. (67). This excess quantity is called the excess free energy. It is the excess free energy of an actual solution when compared with an ideal solution.

2.2. HETEROGENEOUS EQUILIBRIUM—THE PHASE RULE

So far we have only considered homogeneous equilibrium. By this we mean that each system considered has consisted of a single phase, be it a liquid, solid or vapour. In alloy phase equilibria we are mainly concerned with systems composed of two or more phases, whether they be liquid and solid phases, two solid phases, etc. A system in which there is more than one phase is called a heterogeneous system. The subject of heterogeneous equilibrium is our principal concern.

To introduce the concept of heterogeneous equilibrium we will begin by considering a single phase of variable composition. Let there be n_A moles of metal A and n_B moles of metal B in the phase. It is assumed that the phase is in equilibrium at a temperature T and pressure P. The Gibbs free energy of the phase is a function of T, P and the composition of the phase:

$$G' = f(T, P, n_A, n_B).$$

Now imagine that we carry out a reversible process in which the temperature and pressure are changed by dT and dP respectively, and the number of moles of A and B are changed by

dn_A and dn_B. The change in the Gibbs free energy of an arbitrary amount of the phase is then given as

$$dG' = \frac{\partial G'}{\partial P} \cdot dP + \frac{\partial G'}{\partial T} \cdot dT + \frac{\partial G'}{\partial n_A} \cdot dn_A + \frac{\partial G'}{\partial n_B} \cdot dn_B. \tag{68}$$

The prime indicates, as before, an arbitrary amount of the phase.

If dn_A and dn_B were zero (i.e., the mass and the composition of the phase remained constant in the reversible process) then eqn. (68) becomes

$$dG' = \frac{\partial G'}{\partial P} \cdot dP + \frac{\partial G'}{\partial T} \cdot dT$$

or,

$$dG' = V' \, dP - S' \, dT.$$

Hence, eqn. (68) may be re-written as

$$dG' = V' \, dP - S' \, dT + \frac{\partial G'}{\partial n_A} \, dn_A + \frac{\partial G'}{\partial n_B} \, dn_B. \tag{69}$$

The quantity $\partial G'/\partial n_A$ is called the chemical potential of A and is denoted by the symbol μ_A. The chemical potential is equivalent to the partial molar free energy, i.e.,

$$\mu_A = \bar{G}_A; \quad \mu_B = \bar{G}_B; \quad etc.$$

Equation (69) now becomes

$$dG' = V' \, dP - S' \, dT + \mu_A \, dn_A + \mu_B \, dn_B. \tag{70}$$

The chemical potential is a measure of the effect on the Gibbs free energy when a component is introduced into a phase.

From eqn. (34)

$$G' = \mu_A n_A + \mu_B n_B \tag{71}$$

If we considered a gram-atom of the phase rather than an arbitrary amount of the phase, then using eqn. (41)

$$G = \mu_A X_A + \mu_B X_B \tag{72}$$

If there were only one component present in the phase, say metal A, then eqn. (71) would reduce to

$$G' = \mu_A^{\bullet} n_A$$

or,

$$\mu_A^{\bullet} = \frac{G'}{n_A} = G,$$

the symbol $\mu_A^{\bullet}$ being introduced to denote the chemical potential of pure A. The latter is equal to the molar free energy of pure A and is a function of T and P only.

Now consider a heterogeneous system (Fig. 9) containing two phases—an α phase and a β phase. We will also assume that there are only two components in the system—metals A and B. It will be demonstrated that under equilibrium conditions the chemical potential of a component is identical in each phase of a heterogeneous system

i.e.,

$$\mu_A^{\alpha} = \mu_A^{\beta}$$

(the chemical potential of A is the same in the α phase as in the β phase).

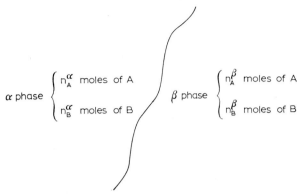

Fig. 9. Heterogeneous equilibrium between an α phase and a β phase in a two-component system.

Using eqn. (70) we can write the free energy equations for the α phase and the β phase as

$$dG'^{\alpha} = V'^{\alpha}\,dP - S'^{\alpha}\,dT + \mu_A^{\alpha}\,dn_A^{\alpha} + \mu_B^{\alpha}\,dn_B^{\alpha} \tag{73}$$
$$dG'^{\beta} = V'^{\beta}\,dP - S'^{\beta}\,dT + \mu_A^{\beta}\,dn_A^{\beta} + \mu_B^{\beta}\,dn_B^{\beta} \tag{74}$$

If we consider the system to be held at constant temperature and pressure then dP and dT are zero. Therefore

$$dG'^{\alpha} = \mu_A^{\alpha}\,dn_A^{\alpha} + \mu_B^{\alpha}\,dn_B^{\alpha} \tag{75}$$
$$dG'^{\beta} = \mu_A^{\beta}\,dn_A^{\beta} + \mu_B^{\beta}\,dn_B^{\beta}. \tag{76}$$

Now imagine that atoms of A are lost by the β phase and transferred to the α phase (Fig. 10). If dn_A moles of A are transferred, the resulting free energy changes of the α and β phases, from eqns. (75) and (76), are

$$dG'^{\alpha} = \mu_A^{\alpha}\,dn_A^{\alpha}$$
$$dG'^{\beta} = \mu_A^{\beta}\,dn_A^{\beta}$$

since for the α phase n_B^{α} does not change, and for the β phase n_B^{β} does not change, *i.e.*, in the transfer of atoms of A $dn_B^{\alpha} = dn_B^{\beta} = 0$.

The total free energy change of the system is equal to the sum of the free energy change of the α and β phases

$$dG' = dG'^{\alpha} + dG'^{\beta}$$
$$= \mu_A^{\alpha}\,dn_A^{\alpha} + \mu_A^{\beta}\,dn_A^{\beta}. \tag{77}$$

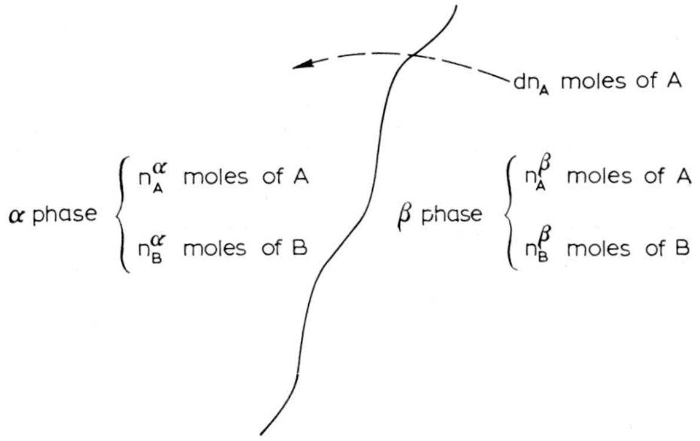

Fig. 10. Transfer of dn_A moles of component A from the β to the α phase.

But $dn_A^{\alpha} = -dn_A^{\beta}$, since the amount of A lost by the β phase (dn_A^{β}) is equal to the amount of A gained by the α phase (dn_A^{α}). Therefore

$$dG' = (\mu_A^{\alpha} - \mu_A^{\beta})\, dn_A^{\alpha}. \tag{78}$$

But $dG' = 0$ for any change under equilibrium conditions at constant T and P in a system of fixed mass and composition. Thus

$$(\mu_A^{\alpha} - \mu_A^{\beta})\, dn_A^{\alpha} = 0$$

and

$$\mu_A^{\alpha} = \mu_A^{\beta} \tag{79}$$

Thus we see that the chemical potential of a component is equal in all the phases of an equilibrium system. In general, for a heterogeneous system:

$$\mu_A^{\alpha} = \mu_A^{\beta} = \mu_A^{\gamma} = \mu_A^{\delta} = \cdots$$
$$\mu_B^{\alpha} = \mu_B^{\beta} = \mu_B^{\gamma} = \mu_B^{\delta} = \cdots$$
$$\mu_C^{\alpha} = \mu_C^{\beta} = \mu_C^{\gamma} = \mu_C^{\delta} = \cdots$$
$$\cdot \qquad \cdot \qquad \cdot \qquad \cdot \qquad \cdots$$
$$\cdot \qquad \cdot \qquad \cdot \qquad \cdot \qquad \cdots$$
$$\cdot \qquad \cdot \qquad \cdot \qquad \cdot \qquad \cdots$$
$$\mu_Z^{\alpha} = \mu_Z^{\beta} = \mu_Z^{\gamma} = \mu_Z^{\delta} = \cdots$$

In deriving the principle of the equality of chemical potential (eqn. (79)), we dealt with an equilibrium system for which $dG' = 0$. If the system were not in equilibrium $dG' < 0$, and hence

$$(\mu_A^{\alpha} - \mu_A^{\beta})\, dn_A^{\alpha} < 0. \tag{80}$$

In this case dn_A^{α} must have the same sign as $(\mu_A^{\beta} - \mu_A^{\alpha})$ to keep the expression in eqn. (80) negative. Thus the chemical potential of A in the β phase is greater than in the α phase and there is therefore a tendency for A atoms to move from the β phase to the α phase under the influence of the potential difference. When sufficient A atoms have been transferred to the α phase $\mu_A^{\beta} = \mu_A^{\alpha}$, and equilibrium will have been reached.

Since a component has identical chemical potentials in each phase of an equilibrium system the chemical potential is a characteristic of the system and not of the phases. There is no need to use the superscript in dealing with chemical potentials, *i.e.* we will use μ_A, μ_B, *etc.* and not μ_A^{α}, μ_B^{α}, *etc.*

A component also has equal activity in each phase of an equilibrium system. Component A (Fig. 9) has an activity a_A^{α} in the α phase, and a_A^{β} in the β phase. Transfer a mole of A from the β phase to the α phase at constant temperature and pressure, and in such a way that there is only an infinitesimal change in the composition of the α and β phases. This can be carried out if we assume that the quantities of the α and β phases are very large. The partial molar free energy of component A in the α phase will be (eqn. 56):

$$\bar{G}_A^{\alpha} = G_A^{\bullet} + RT \ln a_A^{\alpha}$$

or

$$\mu_A^{\alpha} = \mu_A^{\bullet} + RT \ln a_A^{\alpha}. \tag{81}$$

Similarly, for the β phase

$$\mu_A^{\beta} = \mu_A^{\bullet} + RT \ln a_A^{\beta}. \tag{82}$$

The reaction has the form

$$A(\beta) \rightarrow A(\alpha).$$

The free energy change for this reaction will be

$$\Delta G = \mu_A^\alpha - \mu_A^\beta = RT \ln \frac{a_A^\alpha}{a_A^\beta}. \tag{83}$$

If $a_A^\alpha < a_A^\beta$, the logarithm will be negative and so will ΔG. The reaction will proceed until sufficient A has been transferred to the α phase to make $a_A^\alpha = a_A^\beta$. Under these conditions $\Delta G = 0$ and equilibrium has been reached. In equilibrium, therefore, the activity of component A is identical in all phases present in the system.

2.2.1. The phase rule

We have spoken of *components* and *phases* without defining precisely what these terms mean. As we have noted, a heterogeneous system consists of parts of different chemical or physical properties, each part of which is homogeneous. Gibbs* called the homogeneous parts of the system "phases". Each phase in an heterogeneous system is separated from other phases by bounding surfaces. The number of components of a system are the smallest number of independently variable constituents by means of which the composition of each phase involved in the equilibrium may be expressed. In alloy systems the components are normally the metals forming the alloy. An alloy of copper and zinc containing 60 wt. % Cu and 40 wt. % Zn contains two solid phases, the α and β phases, which differ in composition. In this example the components of the system are the metals Cu and Zn, and the system is the mixture of the α and β phases.

Assume that a system in equilibrium has c components distributed between p phases. The composition of each phase is defined by $(c-1)$ concentration terms**. To define the compositions of all p phases it is necessary to use $p(c-1)$ concentration terms. Thus the total number of concentration variables is $p(c-1)$. In addition, we may vary the temperature and pressure of the system. Therefore the total number of variables is $p(c-1)+2$. We will now consider how many variables are defined by the fact that the system is in equilibrium. We already know that in such a system the chemical potential of a component is the same in all phases. For a system containing two components, A and B, and three phases, α, β, and γ, we have:

$$\mu_A^\alpha = \mu_A^\beta = \mu_A^\gamma$$
$$\mu_B^\alpha = \mu_B^\beta = \mu_B^\gamma.$$

Two independent equations determine the equilibrium between the three phases for each component, *i.e.* the equations $\mu_A^\alpha = \mu_A^\beta$ and $\mu_A^\beta = \mu_A^\gamma$, since they yield the third relation $\mu_A^\alpha = \mu_A^\gamma$.

For each component there are $(p-1)$ independent equations relating the chemical potential of that component in all the phases. For the general case with p phases and c components the fact that the system is in equilibrium gives the following set of equations:

---

* *The Scientific Papers of J. Willard Gibbs*, Vol. 1, Dover Publications Inc., 1961, (republication of 1st edn. by Longmans, Green & Co., 1906), p. 96.
** In a three-component system containing A, B and C as components $X_A + X_B + X_C = 1$, and we only need to know two terms to obtain the third by difference.

$$\mu_A^\alpha = \mu_A^\beta = \mu_A^\gamma = \ldots = \mu_A^p$$
$$\mu_B^\alpha = \mu_B^\beta = \mu_B^\gamma = \ldots = \mu_B^p$$

$$\cdot \qquad \cdot \qquad \cdot \qquad \cdots \qquad \cdot$$
$$\cdot \qquad \cdot \qquad \cdot \qquad \cdots \qquad \cdot$$
$$\cdot \qquad \cdot \qquad \cdot \qquad \cdots \qquad \cdot$$

$$\mu_C^\alpha = \mu_C^\beta = \mu_C^\gamma = \ldots = \mu_C^p$$

In this general case there are $c(p-1)$ independent equations. Thus we define or fix $c(p-1)$ variables when we stipulate that the system is in equilibrium.

The number of independent variables or the number of variations which can be made independently is then equal to the total number of variables minus those variables which are automatically fixed.

Number of independent variables
$\qquad$ = total number of variables − number of variables automatically fixed
$$f = [p(c-1)+2] \qquad\qquad - [c(p-1)]$$
$$f = c-p+2. \tag{84}$$

The term f, equal to the number of independent variables, is called the variance of the system or the degrees of freedom of the system. In its usual form eqn. (84) is written

$$p+f = c+2. \tag{85}$$

This relation is known as Gibbs Phase Rule. It defines the condition of equilibrium in a heterogeneous system by the relation between the number of co-existing phases, p, and components, c.

In the derivation of the Phase Rule we have assumed that each phase contains all the components in independently variable proportions. This is not a necessary condition. We can show that eqn. (84) is obtained if one or more components is missing from a particular phase. As an illustrative example we will consider a system with three phases, α, β, and γ, and three components, A, B and C. The α phase contains components A and B only; the β and γ phases contain all three components.

α	β	γ	phases
A	A	A	
B	B	B	components
	C	C	

The number of concentration variables in this case is $(p-1)(c-1)$ for the β and γ phases, and $(c-2)(1)$ for the α phase. A total of $p(c-1)-1$, *i.e.* one less than the previous case where c components were distributed between p phases. The total number of variables is $p(c-1)+1$. The number of variables automatically fixed is obtained from a consideration of the equations of chemical potential:

$$\mu_A^\alpha = \mu_A^\beta = \mu_A^\gamma$$
$$\mu_B^\alpha = \mu_B^\beta = \mu_B^\gamma$$
$$\mu_C^\beta = \mu_C^\gamma.$$

The number of equations which determine the equilibrium for the A and B components are $(c-1)(p-1)$. For the component C the number of independent equations is $(1)(p-2)$. The total number of independent equations is therefore $c(p-1)-1$, which is again one less than that found previously.

The variance

$$f = [p(c-1)+1] - [c(p-1)-1]$$
$$= c - p + 2.$$

Thus we see that the number of variables and the number of equations which determine the equilibrium are reduced by the same amount and we are again led to eqn. (84).

In alloy phase equilibria variations in pressure usually produce no significant effect on the equilibrium, and it is therefore customary to ignore the pressure variable and the vapour phase. In this case one of the variables has been arbitrarily omitted and the Phase Rule for condensed phases (*i.e.*, liquid or solid) only is written

$$p + f = c + 1. \tag{86}$$

As Gibbs pointed out, the Phase Rule as formulated by eqn. (84) is based on the assumption that temperature, pressure and concentration are the only variables which influence the phase relationships. The effect of surface-tension forces at boundaries between phases, of gravitation fields, magnetic fields, etc., is considered to be of negligible importance. Should they assume importance, as for example when considering the solubility of very small particles of one phase in another phase where the energy of the boundary is of relevance, in any particular instance the constant in eqn. (84) would have to be increased by one for each extra variable considered.

By way of introduction to the application of the Phase Rule to alloy phase equilibria it will be instructive to consider systems containing only one component*. Such a system is that formed by a pure metal. In this respect it is not an alloy but the usefulness of this approach will become clear later. Beginning with a pure metal it is obvious that the only variables are the temperature and pressure, since there is no concentration variable. The phase diagram of a pure metal will, therefore, be represented by a graph of P against T. Let us assume that we have solid, liquid, and vapour present as co-existing phases in equilibrium. There are three phases and one component. Applying the Phase Rule,

$$p + f = c + 2$$
$$3 + f = 1 + 2 \quad \text{or} \quad f = 0.$$

There are no degrees of freedom or, in usual parlance, the system is invariant. This statement means that the co-existence of three phases in a unary system can occur at a fixed temperature and pressure only; on the P–T diagram (Fig. 11) the three-phase equilibrium solid/liquid/vapour is represented by a point O.

In addition to the three-phase equilibrium, there are also the following two-phase equilibria:

$$l \rightleftharpoons s \quad \text{co-existing liquid and solid phases}$$
$$l \rightleftharpoons v \quad \text{co-existing liquid and vapour phases}$$
$$s \rightleftharpoons v \quad \text{co-existing solid and vapour phases.}$$

According to the Phase Rule two phases in a unary system have one degree of freedom. This implies that we may arbitrarily choose one of the variables—either pressure or temperature. Having chosen this variable, the other is automatically fixed. Thus the equilibrium between two phases is represented on the P–T diagram by a curve. In Fig. 11 these are the curves OA

---

* A one-component system is called a unary system; a two-component—binary; three-component—ternary; four-component—quaternary; five-component—quinary; six-component—senary.

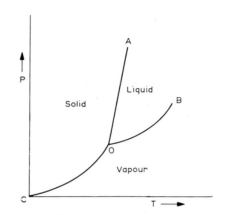

Fig. 11. *P–T* diagram for a unary system (schematic).

(equilibrium between liquid and solid), OB ($l \rightleftharpoons v$) and OC ($s \rightleftharpoons v$). A system with one degree of freedom is said to be monovariant.

Finally, we have three regions which are occupied by a single homogeneous phase—either the liquid, solid or vapour phase. The Phase Rule indicates that a one-phase region in a unary system has two degrees of freedom ($p+f = c+2$; $1+f = 1+2$). We may therefore vary both the temperature and pressure arbitrarily in such a region. If we choose a particular temperature we are also at liberty to select the pressure arbitrarily, provided of course we stay within the boundaries of the region (*i.e.* for the solid phase within the area enclosed by curves OA, OC, and the pressure axis). A system with two degrees of freedom is said to be bivariant.

In this particular case of a pure metal which exists in one solid form only (there are no allotropic modifications) the Phase Rule has defined for us a triple point O at which solid, liquid and vapour are in equilibrium, and from which radiate three curves representing two-phase equilibria. It cannot tell us where point O will lie or in what directions the three curves OA, OB and OC will proceed. However, using the thermodynamic relations previously derived, we can gain further knowledge of this system.

The slope of the monovariant curves OA, OB and OC can be obtained from the Clausius–Clapeyron equation

$$\frac{\mathrm{d}P}{\mathrm{d}T} = \frac{\Delta H}{T \Delta V} \tag{87}$$

where ΔH is the heat absorbed or evolved per mole of metal during the phase change ($l \rightleftharpoons s$, $l \rightleftharpoons v$, or $s \rightleftharpoons v$), ΔV is the change in molar volume for the phase change, and T is the absolute temperature. Taking the liquid–vapour equilibrium we can write the phase change as $l \rightleftharpoons v$, denoting that liquid and vapour are in equilibrium at particular values of P and T. Under these conditions the free energy change for the phase change is zero. By definition

$$\Delta G = G^v - G^l$$

and so

$$G^v = G^l.$$

Differentiating

$$\mathrm{d}G^v = \mathrm{d}G^l.$$

Since (see p. 13) $\mathrm{d}G = V\,\mathrm{d}P - S\,\mathrm{d}T$ (88)

$$dG^v = V^v \, dP - S^v \, dT \tag{89}$$

and

$$dG^l = V^l \, dP - S^l \, dT. \tag{90}$$

Subtracting eqn. (90) from (89),

$$0 = dP(V^v - V^l) - dT(S^v - S^l).$$

Therefore

$$\frac{dP}{dT} = \frac{\Delta S}{\Delta V}.$$

But

$$\Delta S = \frac{\Delta H}{T}.$$

Therefore

$$\frac{dP}{dT} = \frac{\Delta H}{T \Delta V}.$$

In the case considered ΔH and ΔV are the molar enthalpy and volume changes on vapourisation. Since the molar volume of the vapour phase is very much greater than that of either the liquid (or solid) phase we could write

$$\Delta V = V^v - V^l \simeq V^v$$

as an approximation. Considering the vapour as an ideal gas

$$V^v = \frac{RT}{P}.$$

Substituting in eqn. (87) yields

$$\frac{dP}{dT} = \frac{P \Delta H}{RT^2}$$

or,

$$\frac{dP}{P} = \frac{\Delta H}{RT^2} \cdot dT$$

or,

$$\frac{d \ln P}{d(1/T)} = -\frac{\Delta H}{R}. \tag{91}$$

Integrating eqn. (91),

$$\ln P = -\frac{\Delta H}{RT} \tag{92}$$

or,

$$P = e^{-\Delta H/RT}.$$

Equation (92) illustrates the exponential dependence of vapour pressure on temperature, *i.e.*, curves OB and OC have an exponential form. Curve OB is called the vapourisation curve and curve OC the sublimation curve. Equation (92) also illustrates the fact that a plot of $\ln P$ against $1/T$ gives the value $-\Delta H/R$ as the slope of the curve at the temperature concerned.

If we apply the Clausius–Clapeyron equation to the solid–liquid equilibrium, ΔH will be the molar enthalpy change or the molar heat of fusion, and ΔV will be the molar volume change on fusion. The quantity ΔH is always positive but ΔV may be either positive or negative. Therefore the melting curve OA can have a positive or negative slope. Usually the volume of the liquid is greater than the volume of the solid ($V^l - V^s = \Delta V > 0$) and the curve OA has a positive slope, as indicated in Fig. 11. If the reverse were the case, as with the metals bismuth and gallium, curve OA would have a negative slope.

The preceeding consideration of a one-component system illustrates the additional information one can gain by using supplementary thermodynamic relations in conjunction with the Phase Rule.

For most metals the triple point pressure is well below atmospheric pressure. This means that such metals, when heated at atmospheric pressure, melt when the temperature reaches curve OA and then boil when curve OB is reached. Certain metals, such as arsenic, have a triple point well above atmospheric pressure. When heated at atmospheric pressure such metals meet curve OC and are transformed from solid to vapour by sublimation.

2.3. FREE ENERGY OF SOLUTIONS

We have already considered solutions from a thermodynamic point of view, (p. 14, section 2.1). A statistical approach is useful in extending our previous considerations. The simplest model of a solution is one in which the total energy of the solution is given by a summation of interactions between nearest neighbours. In a binary solution with two atom types, A and B, there will be three interaction energy terms

   (1) the energy of A–A pairs of atoms,
   (2) the energy of B–B pairs of atoms,
   (3) the energy of A–B pairs of atoms.

Given the assumption that the total energy of the solution arises from interaction between nearest neighbours only, the total energy of a homogeneous solution A–B will be

$$E = w_{AA}E_{AA} + x_{BB}E_{BB} + y_{AB}E_{AB}$$

where w_{AA}, x_{BB} and y_{AB} are the number of bonds of the type A–A, B–B and A–B respectively and E_{AA}, E_{BB} and E_{AB} are the energies of the bonds A–A, B–B and A–B. The enthalpy of the homogeneous solution can be represented similarly as

$$H = w_{AA}H_{AA} + x_{BB}H_{BB} + y_{AB}H_{AB} \tag{93}$$

where the terms H_{AA}, H_{BB} and H_{AB} are introduced to represent the enthalpies of the bond types. They are again taken to be a measure of the strength of the bonding between the atoms. We require to find w_{AA}, x_{BB} and y_{AB} in order to evaluate H. In order to do this we will make the assumptions that the atoms are arranged on a regular lattice so that each atom has a fixed number of neighbouring atoms (the co-ordination number of the lattice) and that there is a random arrangement of A and B atoms.

If the atomic fraction of A in the homogeneous solution is X_A, then the chance of finding an A atom at any lattice point is also X_A. The chance of finding a B atom at the same lattice point is $(1 - X_A)$ or X_B. The chance of finding two A atoms next to each other is X_A^2; for two B atoms $(1 - X_A)^2$. The probability of finding an A atom next to a B atom or a B atom next to an A atom is $2(X_A)(1 - X_A)$. The factor 2 is introduced since the probability of an A atom being on site 1 and a B atom being on the neighbouring site 2 is $X_A(1 - X_A)$, corresponding to an A–B bond. The probability of the B atom being on site 1 with the A atom on site 2, a B–A bond, is $(1 - X_A)(X_A)$. Therefore the probability that sites 1 and 2 will be occupied by dissimilar atoms is $X_A(1 - X_A) + (1 - X_A)(X_A)$ or $2X_A(1 - X_A)$.

If the solution has a co-ordination number z ($z = 8$ for body-centred cubic lattices and $z = 12$ for face-centred cubic and hexagonal close-packed lattices) and there are a total of N atoms in the solution, then the total number of bonds will be $\frac{1}{2}zN$. The factor $\frac{1}{2}$ arises because each atom shares a bond with its neighbour. The number of A–A bonds will be equal to the total number of bonds multiplied by the probability that a bond will be of the type A–A i.e.,

$$w_{AA} = \tfrac{1}{2}zN(X_A)^2 = \tfrac{1}{2}zNX_A^2$$

similarly

$$x_{BB} = \tfrac{1}{2}zN(1-X_A)^2$$

and

$$y_{AB} = \tfrac{1}{2}zN[2X_A(1-X_A)] = zNX_A(1-X_A).$$

Substituting these values in eqn. (93),

$$H = \tfrac{1}{2}zNX_A^2H_{AA} + \tfrac{1}{2}zN(1-X_A)^2H_{BB} + zNX_A(1-X_A)H_{AB}. \tag{94}$$

Equation (94) expresses the enthalpy of the homogeneous solution. It reduces to the following:

$$H = \frac{zN}{2}\left[X_AH_{AA} + (1-X_A)H_{BB} + 2X_A(1-X_A)\left(H_{AB} - \frac{H_{AA}+H_{BB}}{2}\right)\right]. \tag{95}$$

The enthalpy of pure A and pure B can be obtained from eqn. (94) by substituting the values $X_A = 1$ and $X_A = 0$ respectively. For pure A we find

$$H_A = \frac{zN\,H_{AA}}{2}$$

and for pure B

$$H_B = \frac{zN\,H_{BB}}{2}.$$

A purely mechanical mixture (or an ideal solution) of a fraction X_A of pure A and a fraction $(1-X_A)$ of pure B would have an enthalpy equal to

$$X_AH_A + (1-X_A)H_B = \frac{zN}{2}[X_AH_{AA} + (1-X_A)H_{BB}].$$

This expression is equal to the first two terms on the right of eqn. (95). The third term in eqn. (95) is an additional term arising from the interaction between A and B atoms in the homogeneous solution. It is a measure of the departure of the solution from ideality. Denoting this change in enthalpy on forming a non-ideal homogeneous solution (enthalpy of mixing or enthalpy of formation) by ΔH_m, we see that

$$\Delta H_m = zNX_A(1-X_A)\left[H_{AB} - \frac{H_{AA}+H_{BB}}{2}\right]. \tag{96}$$

Since $zNX_A(1-X_A)$ is always positive the sign of ΔH_m is dependent on that of

$$\left[H_{AB} - \frac{H_{AA}+H_{BB}}{2}\right].$$

Before discussing the implications of this statement it is necessary to recall that the change in free energy on formation of a homogeneous solution is dependent on an entropy effect as well as an enthalpy effect. We have already noted that the entropy of a homogeneous solution is greater than the entropy of a mechanical mixture of the components by an amount $-Nk[X_A \ln X_A + (1 - X_A) \ln (1 - X_A)]$. If the entropy of the system before mixing were S_0 and after mixing S, the change is entropy on mixing, *i.e.* the additional entropy arising on the formation of a homogeneous solution, ΔS_m, is given by

$$S - S_0 = \Delta S_m = -Nk[X_A \ln X_A + (1 - X_A) \ln (1 - X_A)]. \qquad (17)$$

We also noted that the entropy always increases on the formation of a homogeneous solution (*cf.* Fig. 3). The entropy effect can be said to favour the formation of a homogeneous solution with its random arrangement of A and B atoms. The enthalpy may increase or decrease depending on whether H_{AB} is greater or less than $\frac{1}{2}(H_{AA} + H_{BB})$. Since binding energies are all negative, heat always being liberated when a gas is isothermally condensed, the meaning to be attached to such an expression as $H_{AB} > \frac{1}{2}(H_{AA} + H_{BB})$ is that H_{AB} is less negative than $\frac{1}{2}(H_{AA} + H_{BB})$.

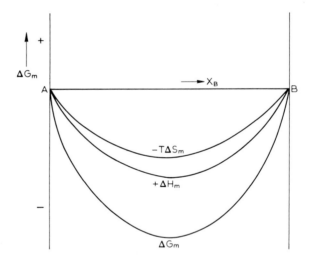

Fig. 12. Variation of free energy with composition for a homogeneous solution with $\Delta H_m < 0$.

Combination of eqns. (96) and (17) allows one to obtain an expression for the change in free energy on formation of a homogeneous solution, *i.e.* the amount by which the free energy of the solution differs from the free energy of a mechanical mixture of the pure components. Since

$$\Delta G_m = \Delta H_m - T\Delta S_m$$

$$\Delta G_m = zNX_A(1 - X_A)\left[H_{AB} - \frac{H_{AA} + H_{BB}}{2}\right] + NkT[X_A \ln X_A + (1 - X_A) \ln (1 - X_A)]. \qquad (97)$$

The shape of the free energy–composition curve at a particular temperature and pressure will depend on the value to be placed on the term $H_{AB} - \frac{1}{2}(H_{AA} + H_{BB})$. If $H_{AB} < \frac{1}{2}(H_{AA} + H_{BB})$ the enthalpy term (eqn. (96)) will always be negative. Since the function $X_A(1 - X_A)$ is parabolic the ΔH_m–composition curve will have the shape illustrated in Fig. 12. The entropy term appears as $-T\Delta S_m$ and is therefore the image of the curve given in Fig. 3. Adding the two curves gives the curve for the change in free energy as a function of composition.

With $H_{AB} < \frac{1}{2}(H_{AA} + H_{BB})$, H_{AB} will be more negative than the mean of $H_{AA} + H_{BB}$. This implies that dissimilar atoms attract more strongly than similar atoms. There will be a tendency towards the formation of a superlattice or a compound.

If $H_{AB} = \frac{1}{2}(H_{AA} + H_{BB})$ the enthalpy term is zero ($\Delta H_m = 0$). These conditions represent the state of affairs for an ideal solution. The enthalpy of such a solution (Fig. 13) is expressed as

$$H = \frac{zN}{2}[X_A H_{AA} + (1 - X_A)H_{BB}].$$

There is no preference in the solution for any particular bond, whether A–A, B–B or A–B. The entropy term is the controlling factor and leads to the formation of a homogeneous solution.

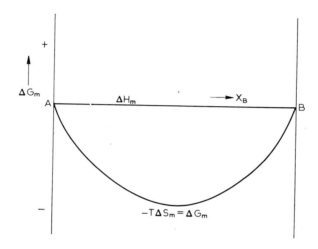

Fig. 13. Variation of free energy with composition for an ideal solution with $\Delta H_m = 0$.

Finally, if $H_{AB} > \frac{1}{2}(H_{AA} + H_{BB})$, ΔH_m will be positive. In the two previous cases there is always a decrease in free energy on the formation of the solution. In this case the enthalpy term is always positive, whereas the entropy term is always negative. As we have already noted (p. 12) the enthalpy (energy) and entropy functions can be considered as a mathematical formulation of two opposing tendencies in a system. To achieve a minimum free energy change, ΔG_m, the system can assume a maximum ΔS_m (and hence a maximum $-T\Delta S_m$) by the destruction of order but under these conditions ΔH_m would not be a minimum. For ΔH_m to be a minimum the system requires to be ordered. With H_{AB} greater than the mean of $H_{AA} + H_{BB}$ similar atoms will attract each other more strongly than dissimilar atoms. There will be a tendency towards the formation of a mechanical mixture of A and B rather than towards the formation of a homogeneous solution. The tendency towards phase separation into phases A and B is dependent on ΔH_m predominating over ΔS_m. With increase in temperature the entropy term becomes more significant and eventually predominates over the enthalpy term. At low temperatures the system will consist of virtually pure A and pure B. As the temperature rises the entropy effect gives rise to increasing association of A and B atoms, i.e. to increasing mutual solubility of A and B. Eventually a temperature will be reached, the critical temperature, at which complete solubility of A and B is attained. The entropy effect is exerting such a predominating role that it completely swamps the enthalpy effect and a homogeneous solution results.

The struggle between the enthalpy term and the entropy term is illustrated in Fig. 14. To explain this figure we note that eqn. (97) can be written in the form

$$\Delta G_m = NCX_A(1-X_A)+NkT[X_A \ln X_A+(1-X_A) \ln (1-X_A)] \tag{98}$$

where

$$C = z \left[H_{AB} - \frac{H_{AA}+H_{BB}}{2}\right].$$

We wish to plot the free energy–composition curves over a range of temperature. In eqn. (98) terms z, N, and X_A are purely numbers, whilst C is an energy term, and k is Boltzmann's constant (energy/temperature). The dimensionless term kT/C can be used as a measure of the temperature of the system. Since N is common to the right-hand side of eqn. (98) we may divide through by N and deal with free energy changes per atom, $\Delta G'_m$, rather than per gram-

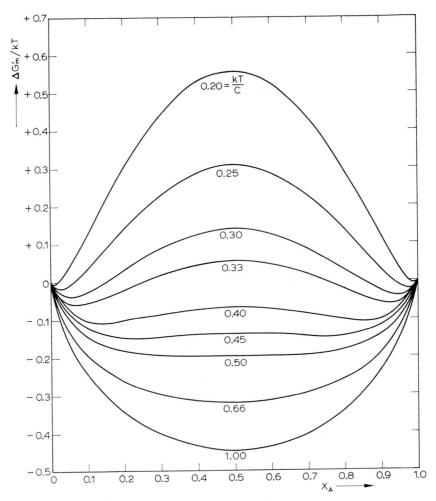

Fig. 14. Variation of free energy with composition for a homogeneous solution with $\Delta H_m > 0$. Free energy–composition curves are given for various values of the parameter kT/C.

atom, ΔG_m. In Fig. 14 are plotted free energy–composition curves for various values of the parameter kT/C. Naturally, low values of the parameter correspond to low temperatures and vice versa. The differing form of the curves is immediately apparent. The curves with $kT/C < 0.5$ show two minima, which approach each other as the temperature rises. Taking the curve $kT/C = 0.25$ as an example, we can substitute $C = 4\,kT$ in eqn. (98) to obtain

$$\frac{\Delta G_m}{N} = \Delta G'_m = kT[4X_A(1-X_A)+X_A \ln X_A+(1-X_A)\ln(1-X_A)].$$

The values of the enthalpy term $4X_A(1-X_A)$ and the entropy term $[X_A \ln X_A+(1-X_A)\ln(1-X_A)]$ are given in Table 1 for a range of X_A values.

Values of X_A above 0.5 are not included since the free energy curve is symmetrical about the vertical axis $X_A = 0.5$. The enthalpy term in column 2 and the entropy term in column 3 are plotted in Fig. 15. Addition of these two curves gives the free energy curve as shown with a minimum at $X_A = 0.02$ and $X_A = 0.98$.

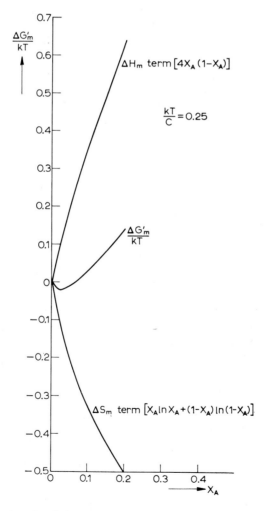

Fig. 15. Summation of enthalpy and entropy effects for a homogeneous solution with $\Delta H_m > 0$ and $kT/C = 0.25$.

TABLE 1

X_A	$4X_A(1-X_A)$	$X_A \ln X_A + (1-X_A) \ln (1-X_A)$	$\Delta G'_m/kT$
0.01	+0.0396	−0.0561	−0.0165
0.015	+0.0591	−0.0780	−0.0189
0.025	+0.0975	−0.1170	−0.0195
0.03	+0.1164	−0.1347	−0.0183
0.05	+0.19	−0.1988	−0.0088
0.1	+0.36	−0.3251	+0.0349
0.2	+0.64	−0.5004	+0.1396
0.3	+0.84	−0.6109	+0.2291
0.4	+0.96	−0.6730	+0.2870
0.5	+1.00	−0.6932	+0.3068

With $kT/C \geq 0.5$ there is a continuous fall in free energy from $X_A = 0$ to $X_A = 0.5$ and from $X_A = 1$ to $X_A = 0.5$. The free energy curve thus assumes the characteristic form one associates with the formation of homogeneous solutions (*cf.* Fig. 12).

2.4. THE FREE ENERGY OF PHASE MIXTURES

When two metals are mixed together they either form a homogeneous solution or they separate into a mixture of phases. Consider an alloy X (Fig. 16) in the binary system A–B to separate at a particular temperature and pressure into two phases, α and β. We will assume that there are N atoms of alloy X and that the fraction of atoms in the α phase is $(1-x)$ and the fraction of atoms in the β phase is x.

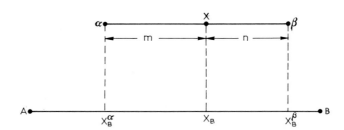

Fig. 16. Separation of alloy X into two phases—α and β.

The number of B atoms in alloy X is taken to be n_B^X, the number of B atoms in the α phase is n_B^α and in the β phase n_B^β. Since the atomic concentration is defined as the number of atoms of one kind divided by the total number of atoms, then

$$X_B = \frac{n_B^X}{N}, \text{ where } X_B \text{ is the atomic concentration of B in alloy } X$$

$$X_B^\alpha = \frac{n_B^\alpha}{N(1-x)}, \text{ where } X_B^\alpha \text{ is the atomic concentration of B in the } \alpha \text{ phase}$$

$$X_B^\beta = \frac{n_B^\beta}{Nx}, \text{ where } X_B^\beta \text{ is the atomic concentration of B in the } \beta \text{ phase.}$$

Since

$$n_B^X = n_B^\alpha + n_B^\beta$$

then

$$NX_B = NX_B^\alpha(1-x) + NX_B^\beta x$$

whence

$$x = \frac{X_B - X_B^\alpha}{X_B^\beta - X_B^\alpha} = \frac{m}{m+n}$$

and

$$1 - x = \frac{X_B^\beta - X_B}{X_B^\beta - X_B^\alpha} = \frac{n}{m+n}$$

and

$$\frac{x}{1-x} = \frac{m}{n}. \tag{99}$$

Equation (99) is called the lever rule. It enables one to calculate the relative amounts of phases in a phase mixture in terms of the composition of the alloy and the phases into which it separates. If weight fractions were used instead of atomic fractions then

$$\frac{m_\beta}{m_\alpha} = \frac{m}{n}$$

or

$$m_\beta n = m_\alpha m.$$

This equation also expresses the condition for the balance of a lever with a fulcrum at X (Fig. 17). Hence the designation "lever rule".

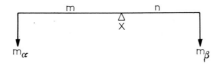

Fig. 17. Balance of a lever with fulcrum at X.

The free energy of a phase mixture can be obtained by utilising the lever rule. If alloy X separates into two phases α and β, the free energy of the alloy will be unchanged by the separation. The free energy of alloy X is equal to the sum of the free energies of the α and β phases. Since alloy X consists of an amount of α phase equal to $(X_B^\beta - X_B)/(X_B^\beta - X_B^\alpha)$ and an amount of β phase $(X_B - X_B^\alpha)/(X_B^\beta - X_B^\alpha)$, the free energy of alloy X will be equal to

$$a\left(\frac{X_B^\beta - X_B}{X_B^\beta - X_B^\alpha}\right) + b\left(\frac{X_B - X_B^\alpha}{X_B^\beta - X_B^\alpha}\right)$$

where a and b are the free energies of the α and the β phase respectively at the temperature and pressure given (Fig. 18). Re-arranging, the free energy of alloy X becomes

$$= a\left(\frac{(X_B^\beta - X_B^\alpha) - (X_B - X_B^\alpha)}{X_B^\beta - X_B^\alpha}\right) + b\left(\frac{X_B - X_B^\alpha}{X_B^\beta - X_B^\alpha}\right)$$

$$= a + (b - a)\left(\frac{X_B - X_B^\alpha}{X_B^\beta - X_B^\alpha}\right)$$

$$= a + c.$$

We see therefore that an alloy X which separates into two phases of composition X_B^α and X_B^β

whose free energies are a and b itself has a free energy which is given by the point X on the straight line $\alpha\beta$. Note that the line $\alpha X\beta$ (Fig. 16) was drawn horizontal to represent isothermal–isobaric conditions (constant temperature and pressure), whereas the line $\alpha X\beta$ (Fig. 18) was given a definite slope to indicate that the free energies of different phases are not identical $(a \neq b)$.

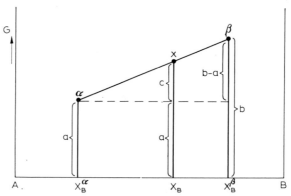

Fig. 18. Free energy of a phase mixture.

We can now return to a consideration of the significance of the free energy curves in Fig. 14. For $kT/C < 0.5$ the curves show two minima. The arbitrary free energy–composition curve (Fig. 19) is drawn without the symmetry of Fig. 14. The free energies of pure A and B, $G_A^\bullet$ and $G_B^\bullet$, at a temperature T will be different, as can be seen from eqn. (26), and the solution formed by A and B does not behave ideally. We will now show that, in this case, a homogeneous solution is not the most stable state of the system. In fact, over a wide range of concentration, the most stable state corresponds to a mixture of phases. In Fig. 19, if alloy X existed as a homogeneous solution its free energy would be represented by point C. However, the free energy can be reduced if alloy X separates into two phases. For example, if the two phases were represented by points D and E the free energy of the phase mixture would be reduced from C to F. The maximum reduction in free energy is obtained when the alloy X separates

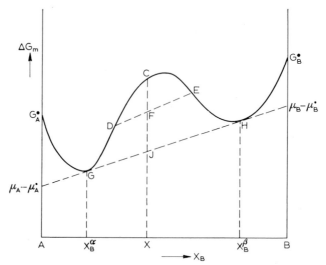

Fig. 19. Asymmetrical free energy–composition curve for $kT/C < 0.5$.

into phases G and H. The free energy of such a phase mixture is given by point J. Points G and H do not correspond with the minima in the free energy curve but to points of contact of the double tangent line to the free energy–composition curve. In the case of a symmetrical free energy curve, such as was illustrated in Fig. 14, the points G and H would coincide with the minima.

It may seem puzzling at first sight that the alloy X separates into two phases with free energies given by G and H respectively, where G may not be the minimum point on the free energy curve. The important criterion for equilibrium, however, is that the chemical potential of a given component is the same in both phases, not that the free energy of each phase be a minimum. The tangent to points G and H can be extrapolated to cut the vertical axis B. The intersection gives the value for $\mu_B - \mu_B^\bullet$, the partial molar free energy of B relative to pure B. It follows that μ_B is the same for both phases into which the alloy X separates. Hence the chemical potential of component B is identical in both phases (*cf.* p. 22).

Any alloy with a composition between X_B^α and X_B^β will separate into two phases, G and H. Its free energy will be represented by a point on the line GH. The proportions of the phases G and H will vary with the alloy composition between G and H, but the compositions of the co-existing phases will be fixed at X_B^α and X_B^β. Alloys with compositions between pure A and X_B^α (and between pure B and X_B^β) are most stable as homogeneous solutions. The compositions of the stable phase or solution is identical to the alloy composition. As the alloy composition is changed from A to X_B^α or from B to X_B^β so too does the phase composition change.

The points G and H represent the limits of solubility of the components. Point G, corresponding to composition X_B^α, represents the amount of B which is soluble in A at the particular temperature and pressure concerned. Point H represents the amount of A which is soluble in B under these conditions.

In the free energy–composition diagrams we have plotted the change in free energy on forming one gram-atom of solution. This quantity is equal to the free energy of the solution minus the free energy of a mechanical mixture of the components. Thermodynamically,

$$\Delta G_m = G - (G_A^\bullet X_A + G_B^\bullet X_B).$$

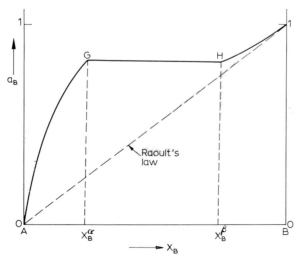

Fig. 20. Activity–composition curve for component B as derived from Fig. 19.

We have already noted that the intercepts, at $X_A = 0$ and 1, of any tangent to a free energy curve give the corresponding values of the partial molar free energies or chemical potentials (*cf.* p. 15). Considering component B, the quantity $\mu_B - \mu_B^\bullet$ can be obtained for any composition between A and B (Fig. 19) by extrapolating the tangent to the free energy curve at that composition to $X_B = 1$. Knowing $\mu_B - \mu_B^\bullet$, and using eqn. (56):

$$\bar{G}_B - G_B^\bullet = \mu_B - \mu_B^\bullet = RT \ln a_B$$

we can directly obtain the activity of component B as a function of composition. The activity–composition curve for component B, as derived from Fig. 19, is given in Fig. 20.

The activity of component B is seen to remain constant over the composition range from X_B^α to X_B^β, complying with the statement on p. 25 that at equilibrium the activity of a component is identical in all phases co-existing in the system.

Finally we may note that Fig. 20 is in reality an extension of Fig. 7c to the case where $A \leftrightarrow B \ll \frac{1}{2}(A \leftrightarrow A + B \leftrightarrow B)$. The attractive force between dissimilar atoms is so much smaller than the attractive force between similar atoms that actual phase separation occurs.

2.5. VARIATION OF SOLUBILITY WITH TEMPERATURE

We have already noted that points G and H (Fig. 19) give the limits of solubility of the components at the temperature and pressure concerned. For the symmetrical case illustrated in Fig. 14 the slope of the ΔG_m–X_A curves are zero at the solubility limits. Differentiating eqn. (98) with respect to X_A and equating to zero, we find

$$\frac{d(\Delta G_m)}{dX_A} = NC(1 - 2X_A) + NkT[\ln X_A - \ln(1 - X_A)] = 0$$

$$\ln\left(\frac{X_A}{1 - X_A}\right) = -\frac{C(1 - 2X_A)}{kT}$$

or

$$T = \frac{C(1 - 2X_A)}{k \ln\left(\frac{1 - X_A}{X_A}\right)}. \tag{100}$$

Figure 21 gives the solubility curve corresponding to eqn. (100). The solubility of the components in each other increases with temperature until a temperature is reached where the components are completely miscible (soluble) in each other. The temperature at which complete miscibility occurs is called the critical temperature, T_c. At T_c the term $d^2(\Delta G_m)/d(X_A)^2$ will be zero. Since

$$\frac{d^2(\Delta G_m)}{d(X_A)^2} = -2NC + NkT_c\left(\frac{1}{X_A} + \frac{1}{1 - X_A}\right) = 0$$

then

$$2C = \frac{kT_c}{X_A(1 - X_A)} \quad \text{or} \quad T_c = \frac{2CX_A(1 - X_A)}{k}$$

The term T_c will be a maximum when $X_A = (1 - X_A) = 0.5$. It follows that

$$T_c = \frac{C}{2k}. \tag{101}$$

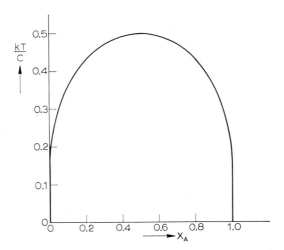

Fig. 21. Solubility curve obtained from eqn. (100) by plotting the concentration of A in two co-existing phases as a function of kT/C.

A high value of the critical temperature is associated with a high positive value for C ($= z[H_{AB} - \frac{1}{2}(H_{AA} + H_{BB})]$).

The stronger the attraction between similar atoms, the higher T_c. In those binary phase diagrams with a miscibility gap in the solid state the gap has not the symmetrical form shown in Fig. 21. This is primarily because the initial simplifying assumption that the energy is the sum of interaction between pairs of atoms is never absolutely valid. The systems Pd–Ir*, Pt–Ir** and Pt–Au*** all have miscibility gaps in the solid state with varying degrees of asymmetry. Most binary phase diagrams with a positive value of ΔH_m do not show a miscibility gap with a closure at temperature T_c since melting occurs before T_c is reached (for example the Ag–Cu system). For attempts to find models for assymmetric miscibility gaps the reader is referred to papers by H. K. HARDY (*Acta Met.*, **1** (1953) 202) and L. J. VAN DER TOORN AND T. J. TIEDEMA (*Acta Met.*, **8** (1960) 711).

---
* E. RAUB AND W. PLATE, *Z. Metallk.*, **48** (1957) 444.
** E. RAUB AND W. PLATE, *Z. Metallk.*, **47** (1956) 688.
*** A. S. DARLING, R. A. MINTERN AND J. C. CHASTON, *J. Inst. Metals*, **81** (1952–53) 125.

CHAPTER 3

Binary Phase Diagrams.
Two-Phase Equilibrium

3.1. INTRODUCTION

When discussing the free energy of solutions and phase mixtures we noted that the free energy–composition curves gave the compositions of co-existing phases under equilibrium conditions. By plotting a series of such curves at different temperatures we established the manner in which the phase compositions changed with temperature. In other words, we determined the phase limits or phase boundaries as a function of temperature. A phase diagram is nothing more than a presentation of data on the position of phase boundaries as a function of temperature.

In this chapter we will make use of two main thermodynamic principles—the Phase Rule and the free energy–composition curves—in order to explain the equilibria in binary alloy systems. The Phase Rule is basic to any consideration of alloy equilibria. Free energy–composition curves can only be used to illustrate the general features of alloy systems since quantitative data is generally lacking and theory inadequate to allow a calculation of the free energy of alloy phases.

According to the condensed Phase Rule, eqn. (86),

$$p+f = c+1.$$

For a binary system the equilibria possible are summarised below.

Number of components	Number of phases	Variance	Equilibrium
$c = 2$	$p = 1$	$f = 2$	bivariant
$c = 2$	$p = 2$	$f = 1$	monovariant
$c = 2$	$p = 3$	$f = 0$	invariant

Bivariant equilibrium : $p = c-1$

A binary system with one phase only has two degrees of freedom. This implies that the temperature and the concentration of one component in the phase (which defines the composition of the binary phase since $X_A = 1-X_B$) may be selected arbitrarily. One-phase equilibrium corresponds to the existence of distinct liquid ($l_1, l_2, \ldots$) or solid ($\alpha, \beta, \ldots$) phase regions in the binary phase diagram.

Monovariant equilibrium : $p = c$

When two phases co-exist in a binary system, the system has one degree of freedom. One of the external variables can be chosen freely but, once this variable has been selected, the other variables are automatically fixed. If the temperature of the system is chosen, then the composition of the co-existing phases will be fixed. Alternatively, if the composition of one phase is chosen by stating the concentration of one component in that phase, then the temperature and the composition of the second phase will be fixed.

The monovariant reactions which are observed in binary systems are: $l \rightleftharpoons \alpha$, $l_1 \rightleftharpoons l_2$, and $\alpha \rightleftharpoons \beta$.

Invariant equilibrium : $p = c+1$

If three phases co-exist in a binary system, there are zero degrees of freedom. The system can therefore only exist at one temperature and the composition of all three phases is fixed.

If heat is applied to a system in invariant equilibrium, the temperature and the compositions of the co-existing phases initially remain constant. A change occurs, however, in the relative amounts of the three phases until eventually one phase is consumed and the system becomes monovariant. (In some cases two of the phases may be consumed simultaneously and the system then becomes bivariant.) This process is analogous to the invariant equilibrium at the melting point of a pure metal. At the melting point liquid is in equilibrium with solid. Addition of heat has the initial effect of increasing the amount of liquid at the expense of the solid phase, until finally all the solid phase is consumed.

The invariant reactions which have been observed in binary diagrams are listed below, together with the nomenclature given to such reactions.

$l \rightleftharpoons \alpha + \beta$	eutectic reaction	(*e.g.* Ag–Cu system)
$\gamma \rightleftharpoons \alpha + \beta$	eutectoid reaction	(*e.g.* C–Fe system)
$l_1 \rightleftharpoons \alpha + l_2$	monotectic reaction	(*e.g.* Cu–Pb system)
$\alpha \rightleftharpoons \beta + l$	metatectic reaction	(*e.g.* Ag–Li system)
$l + \alpha \rightleftharpoons \beta$	peritectic reaction	(*e.g.* Cu–Zn system)
$\alpha + \beta \rightleftharpoons \gamma$	peritectoid reaction	(*e.g.* Al–Cu system)
$l_1 + l_2 \rightleftharpoons \alpha$	syntectic reaction	(*e.g.* K–Zn system)

The symbols l, l_1 and l_2 denote liquid phases, and α, β and γ solid phases. The liquid phase in alloy systems is referred to as either liquid or melt. It should be noted that invariant reactions involving liquid phases have a name ending in *-tectic* whilst those occurring completely in the solid state end in *-tectoid*.

3.2. TWO-PHASE EQUILIBRIUM.

COMPLETE SOLUBILITY OF THE COMPONENTS IN BOTH THE LIQUID AND SOLID STATE

3.2.1. *Thermodynamics*

The simplest case of a binary system is that where components A and B are soluble in all proportions in both the liquid and solid state. This occurs when A and B have very similar atomic radii, the same valency and crystal structure*. Under these conditions the term $H_{AB} - \frac{1}{2}(H_{AA} + H_{BB})$ will be small. To simplify matters we can assume that A and B form ideal solutions in both the liquid and solid state (*i.e.* assume $H_{AB} = \frac{1}{2}(H_{AA} + H_{BB})$). At first sight it would seem that the boundary between the liquid phase and the solid phase should be

---

* For a detailed exposition of structural principles of alloy systems the reader is referred to W. HUME-ROTHERY AND G. V. RAYNOR, *The Structure of Metals and Alloys*, Monograph No. 1, 4th edn., Institute of Metals, London, 1962, and W. HUME-ROTHERY, *Elements of Structural Metallurgy*, Monograph No. 26, Institute of Metals, London, 1961.

a straight line drawn from the melting point of A to that of B (Fig. 22). The only equilibrium possible in the system is that between the liquid phase, l, and the solid phase, $\alpha : l \rightleftharpoons \alpha$. It will now be shown that there is a two-phase, $l+\alpha$, region between the liquid phase and the solid phase regions.

Taking eqn. (70) and re-writing it to refer to one mole of solution, we obtain the change in free energy for the solid and liquid phases as:

$$dG^s = V^s\,dP - S^s\,dT + \mu_A^s\,dX_A^s + \mu_B^s\,dX_B^s$$
$$dG^l = V^l\,dP - S^l\,dT + \mu_A^l\,dX_A^l + \mu_B^l\,dX_B^l.$$

At equilibrium

$$dG^s = dG^l, \quad \mu_A^s = \mu_A^l \quad \text{and} \quad \mu_B^s = \mu_B^l.$$

At constant P

$$-(S^s - S^l)\,dT = (\mu_A - \mu_B)(dX_A^l - dX_A^s).$$

Differentiating with respect to X_A,

$$(S^s - S^l)\frac{dT}{dX_A} = (\mu_A - \mu_B)\left(\frac{dX_A^s}{dX_A} - \frac{dX_A^l}{dX_A}\right). \tag{102}$$

If the boundary between liquid and solid were as shown in Fig. 22, then $X_A^s = X_A^l$ (the liquid and solid would have the same composition when in equilibrium at a point on the line $T_A T_B$). From eqn. (102)

$$(S^s - S^l)\frac{dT}{dX_A} = 0.$$

Since $S^s \neq S^l$, then $dT/dX_A = 0$. Thus the condition $X_A^s = X_A^l$ is only associated with $dT/dX_A = 0$, i.e. with a minimum or a maximum in the line $T_A T_B$ of Fig. 22. Except for this particular case therefore $X_A^s \neq X_A^l$. There is a difference between the composition of the liquid and solid phase in the general case.

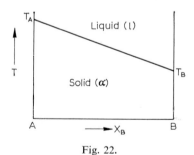

Fig. 22.

In order to derive the form of the phase diagram in a qualitative manner use can be made of the free energy–composition curves. Since the only possible equilibrium is between the liquid and solid α, we require to consider the relative dispositions of the free energy curves for both liquid and α. At temperatures above T_A—the melting point of A—all alloys will be liquid and hence the free energy curve of the α phase will lie above that of the liquid phase (Fig. 23a). When temperature T_A is reached, the free energy curves for liquid and α intersect at a point on the A ordinate. All alloys are stable as a liquid phase (Fig. 23b). With a fall in temperature to T_2 the α curve intersects the liquid curve, as shown in Fig. 23c. From free energy considerations it is seen that alloys with compositions $<X_B^s$ are stable as α solid solution; alloys between X_B^s and X_B^l are stable as a phase mixture of liquid and α; and alloys with compositions $>X_B^l$ are stable as liquid solution.

As the temperature falls further, the point of intersection of the free energy curves moves towards the B ordinate until, at T_B, the curves intersect at pure B. All alloys are solid α phase at T_B; only pure B exists as a liquid in equilibrium with solid. Below T_B all compositions are solid, as can be seen from the disposition of curves in Fig. 23e.

A series of free energy curves such as Fig. 23b–d allow the compositions of the co-existing liquid and solid phases to be plotted. The temperature–concentration diagram so obtained is given in Fig. 23f. On the assumption that both the liquid and solid solutions are ideal, it is

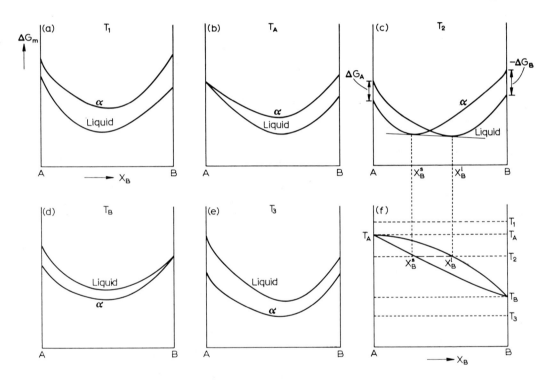

Fig. 23. Derivation of the phase diagram from the free energy curves for the liquid and α phases.
$T_1 > T_A > T_2 > T_B > T_3$.

possible to calculate the compositions of the liquid and solid phases in equilibrium at each temperature between T_A and T_B and thus to derive the diagram shown in Fig. 23f*. This has been done for the Ge–Si system**. Consider the free energy curves for liquid and α at a temperature T, where $T_A > T > T_B$. The standard states are pure *solid* A and pure *liquid* B at temperature T. We require to derive the free energy curves for the liquid and α phases (Fig. 23c).

---

* The treatment of R. A. SWALIN (*Thermodynamics of Solids*, Wiley, New York, 1962, p. 167) is followed.
** C. D. THURMOND, *J. Phys. Chem.*, **57** (1953) 827.

Free energy curve for the liquid phase

For an ideal solution with a reaction

X_A moles of liquid A $+ X_B$ moles of liquid B = 1 mole liquid solution

$$\Delta G_m = RT(X_A^l \ln X_A^l + X_B^l \ln X_B^l). \tag{61}$$

But the standard state for A is pure solid A. To obtain the free energy change, ΔG_m^l, for the formation of the liquid solution, we must allow for the conversion of X_A moles of solid A into X_A moles of liquid A, *i.e.*

X_A moles solid A $= X_A$ moles liquid A

$$\Delta G = X_A^l \Delta G_A$$

where ΔG_A is the difference in free energy between liquid and solid A (the free energy of fusion of A).

Hence for the reaction

X_A moles of solid A $+ X_B$ moles of liquid B = 1 mole liquid solution

$$\Delta G_m^l = X_A^l \Delta G_A + RT(X_A^l \ln X_A^l + X_B^l \ln X_B^l).$$

Free energy curve for the α phase

For an ideal solid solution formed according to the equation

X_A moles of solid A $+ X_B$ moles of solid B = 1 mole α solid solution

$$\Delta G_m = RT(X_A^s \ln X_A^s + X_B^s \ln X_B^s).$$

Since the standard state for B is pure *liquid* B, allowance must be made for the free energy change on transforming X_B moles of liquid B to X_B moles of solid B, *i.e.*

$$\Delta G = -X_B^s \Delta G_B$$

where ΔG_B is the difference in free energy between liquid and solid B. It is a negative quantity since the liquid phase is the more stable at temperature T.

For the reaction

X_A moles of solid A $+ X_B$ moles of liquid B = 1 mole α solid solution

$$\Delta G_m^s = -X_B^s \Delta G_B + RT(X_A^s \ln X_A^s + X_B^s \ln X_B^s).$$

At temperature T the chemical potential of component B is identical in the α and liquid phases. Therefore

$$\frac{\partial \Delta G_m^s}{\partial X_B} = \frac{\partial \Delta G_m^l}{\partial X_B}$$

$$\frac{\partial \Delta G_m^s}{\partial X_B} = -\Delta G_B + RT \ln \frac{X_B^s}{X_A^s}$$

and

$$\frac{\partial \Delta G_m^l}{\partial X_B} = -\Delta G_A + RT \ln \frac{X_B^l}{X_A^l}.$$

Therefore

$$\Delta G_A - \Delta G_B = RT \left(\ln \frac{X_B^l}{X_A^l} - \ln \frac{X_B^s}{X_A^s} \right)$$

or,

$$\Delta G_A - \Delta G_B = RT\left(\ln\frac{X_A^s}{X_A^l} - \ln\frac{X_B^s}{X_B^l}\right).$$

We can relate ΔG_A and ΔG_B to ΔH_A and ΔH_B, T_A and T_B, i.e. to the respective molar heats of fusion of pure A and B, and their melting points. For example, for component B at T_B:

$$\Delta G_B = \Delta H_B - T_B\Delta S_B = 0$$

or,

$$\Delta S_B = \frac{\Delta H_B}{T_B}.$$

At a temperature T,

$$\Delta G_B = \Delta H_B - T\Delta S_B \neq 0.$$

Therefore

$$\Delta G_B = \Delta H_B - T\frac{\Delta H_B}{T_B}$$

$$= \Delta H_B\left(1 - \frac{T}{T_B}\right).$$

Substituting,

$$\Delta H_A\left(1 - \frac{T}{T_A}\right) - \Delta H_B\left(1 - \frac{T}{T_B}\right) = RT\left(\ln\frac{X_A^s}{X_A^l} - \ln\frac{X_B^s}{X_B^l}\right)$$

or,

$$\ln\frac{X_A^s}{X_A^l} - \ln\frac{X_B^s}{X_B^l} = \frac{\Delta H_A}{R}\left(\frac{1}{T} - \frac{1}{T_A}\right) - \frac{\Delta H_B}{R}\left(\frac{1}{T} - \frac{1}{T_B}\right).$$

As the temperature approaches T_A the quantities X_A^s and X_A^l will approach unity, and $1/T$ will approach $1/T_A$.

Hence near T_A:

$$\ln\frac{X_B^s}{X_B^l} = \frac{\Delta H_B}{R}\left(\frac{1}{T} - \frac{1}{T_B}\right). \tag{103}$$

Similarly, if the temperature approaches T_B, $X_B^s \simeq X_B^l \to 1$ and $1/T \to 1/T_B$. Near T_B:

$$\ln\frac{X_A^s}{X_A^l} = \frac{\Delta H_A}{R}\left(\frac{1}{T} - \frac{1}{T_A}\right). \tag{104}$$

Knowing ΔH_A, ΔH_B, T_A and T_B, the above two equations can be used to determine the compositions of co-existing phases at a series of temperatures, T, between T_A and T_B. The resulting phase diagram (Fig. 23f) is shown in Fig. 24 with the region of existence of the various phases included.

In the derivation given above it was assumed that both the liquid and solid solutions were ideal in behaviour. We will now turn to a consideration of actual (or so-called regular) solutions in which $\Delta H_m \neq 0$, but $\Delta S_m = \Delta S_{m,\text{ideal}}$. The treatment follows that of Rand and Kubaschewski* and is based on a calculation of free energy curves for the solid and liquid phases at a

---

* M. H. RAND AND O. KUBASCHEWSKI, *The Thermochemical Properties of Uranium Compounds*, Oliver & Boyd, Edinburgh, 1963, p. 89.

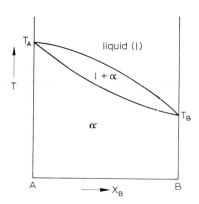

Fig. 24. Phase diagram for complete solubility of components A and B in both the liquid and solid state.

series of temperatures. From such curves the compositions of the co-existing liquid and solid phases at each temperature are obtained by application of the common tangent construction (*cf.* Fig. 23c).

As before, consider the free energy curves for the liquid and solid solutions at a temperature T, where $T_A > T > T_B$. The standard states are taken to be pure solid A and pure liquid B at a temperature T. The free energy curve for the liquid phase is:

$$\Delta G_m^l = \Delta H_m^l + X_A^l \Delta G_A + RT(X_A^l \ln X_A^l + X_B^l \ln X_B^l).$$

Since

$$\Delta G_A = \Delta H_A - T \Delta S_A$$

then,

$$\Delta G_m^l = \Delta H_m^l + X_A^l \Delta H_A - X_A^l T \Delta S_A + RT(X_A^l \ln X_A^l + X_B^l \ln X_B^l).$$

The free energy curve for the solid phase is:

$$\Delta G_m^s = \Delta H_m^s - X_B^s \Delta G_B + RT(X_A^s \ln X_A^s + X_B^s \ln X_B^s)$$

or,

$$\Delta G_m^s = \Delta H_m^s - X_B^s \Delta H_B + X_B^s T \Delta S_B + RT(X_A^s \ln X_A^s + X_B^s \ln X_B^s).$$

Rand and Kubaschewski have illustrated the application of thermodynamics to the calculation of the solidus* and liquidus* in the U–Zr system. With reference to Fig. 23f, component A is now Zr and component B is U. It was assumed that $\Delta H_m^l = \Delta H_m^s$ and the concentration dependence of ΔH_m was given as:

X_B	0.1	0.2	0.3	0.4	0.5	0.6	0.7	0.8	0.9
ΔH_m	340	590	800	940	1010	1000	910	740	468

Heats and entropies of fusion of A and B were taken to be

$$\Delta H_A = \Delta H_{Zr} = 4600 \text{ cal/g.-atom,}$$
$$\Delta H_B = \Delta H_U = 2500 \text{ cal/g.-atom,}$$
$$\Delta S_A = \Delta S_{Zr} = 2.16 \text{ cal/degree,}$$
$$\Delta S_B = \Delta S_U = 1.78 \text{ cal/degree.}$$

Taking a temperature of 1800 °K, we can now calculate ΔG_m^l and ΔG_m^s for various values of $X_B(= X_U)$. The result is given in the following table and graphically in Fig. 25.

---

* These terms are defined on p. 50.

X_B	0.1	0.2	0.3	0.4	0.5	0.6	0.7	0.8	0.9
ΔG_m^l	−189	−642	−901	−1056	−1130	−1138	−1075	−919	−631
ΔG_m^s	−760	−1071	−1189	−1203	−1136	−1003	−799	−499	−72

The common tangent to the free energy curves touches the liquid curve at $X_B = X_U = 0.6$, and the solid curve at $X_B = X_U = 0.4$. At 1800 °K a liquid of composition $X_U = 0.6$ (*i.e.* an atomic fraction of U = 0.6 or an atomic % U = 60) is in equilibrium with a solid of composition $X_U = 0.4$ (an atomic % U = 40). Free energy curves can be calculated for other temperatures between T_{Zr} and T_U. From the solid and liquid compositions so obtained the complete solidus and liquidus can be derived. Figure 26 is a schematic representation of free energy curves at temperatures of 1500, 1800, and 2000 °K. It should be noted that the slope of the common tangent line changes from a high positive slope at 1500 °K to a low positive slope at 1800 °K and to a high negative slope at 2000 °K. The change in slope results from a change in relative dispositions of the liquid and solid free energy curves as the temperature rises. At 1500 °K the free energy curve for the solid phase is below the free energy curve for the liquid

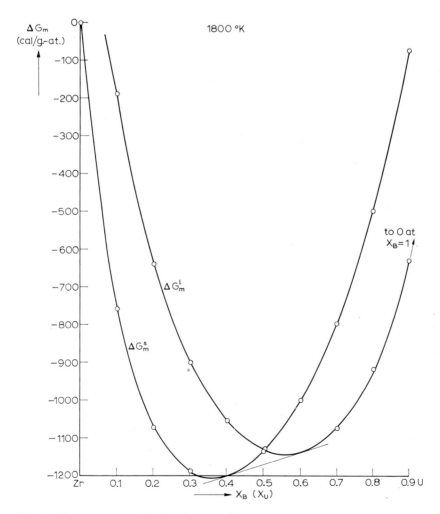

Fig. 25. Free energy curves for liquid and solid phases in the U–Zr system at 1800 °K.
(After M. H. RAND AND O. KUBASCHEWSKI; courtesy Oliver & Boyd.)

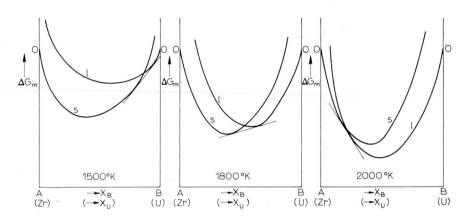

Fig. 26. Free energy curves for liquid and solid phases in the U–Zr system at 1500°, 1800° and 2000 °K.

phase over nearly the whole composition range. As the temperature rises, the liquid free energy curve falls relative to that of the solid (but, as mentioned on p. 12, both free energy curves fall with rise in temperature) until at 2000 °K only a small composition range near component A ($=$ Zr) is stable as the solid phase.

The upper curve on the phase diagram (Fig. 24), originating at T_A and ending at T_B, defines the limits of existence of the liquid phase. It is called the liquidus. Above the liquidus all alloys in the system are liquid. The lower curve defines the limits of existence of the α phase. It is called the solidus. Below the solidus all alloys are solid. Between the liquidus and solidus the alloys consist of liquid and α phase. As we have seen, the compositions of the co-existing liquid and α phases at a particular temperature are given by the intersection of a horizontal line (an isotherm) with the liquidus and solidus respectively. In Fig. 23f at a temperature T_2, liquid of composition X_B^l is in equilibrium with α of composition X_B^s. X_B^l is a point on the liquidus and X_B^s is a point on the solidus. The line $X_B^s X_B^l$, joining the compositions of co-existing phases in a binary system, is called a *tie line**.

Referring to Fig. 23f, if A is regarded as the solvent, for very dilute solutions of B in A we can write

$$X_A \rightarrow 1 \quad \text{and} \quad -\ln X_A \simeq X_B.$$

In terms of eqn. (104):

$$X_A^l - X_A^s = \frac{\Delta H_A}{R}\left(\frac{T_A - T}{T T_A}\right).$$

Since $X_A^l = 1 - X_B^l$ and $X_A^s = 1 - X_B^s$

$$X_B^s - X_B^l = \frac{\Delta H_A}{R}\left(\frac{T_A - T}{T T_A}\right). \tag{105}$$

As T approaches T_A (in dilute solutions of B in solvent A), the denominator on the right-hand side of eqn. (105) can be written RT_A^2. Therefore

$$X_B^s - X_B^l = \frac{\Delta H_A}{R T_A^2}(T_A - T) \tag{106}$$

---

* In German literature the term *konode* is used. It is frequently translated as "conode". The English word "tie line" is preferred.

or,

$$\left(\frac{dX_B^s}{dT} - \frac{dX_B^l}{dT}\right)_{T=T_A} = \frac{\Delta H_A}{RT_A^2}.$$

(107)

Equations (106) and (107) are referred to as the Van't Hoff relation. They give the depression of the freezing point for a liquid solution in equilibrium with a solid solution. The difference in initial slopes of the solidus and liquidus curves, the slopes at $T = T_A$ and $X_A = 1$, are dependent on the latent heat of fusion of pure A (ΔH_A) but independent of the nature of the solute. Van 't Hoff's relation applies only to very dilute solutions, and it is not possible to predict the limits of solute concentration to which it applies.

3.2.2. *Equilibrium freezing of alloys*

It has already been noted that the free energies of the liquid and solid phases are equal at the melting point of a pure metal. Below the melting point the solid phase has the lower free energy and is therefore the stable phase. Referring to Fig. 4, there is a decrease in free energy for the liquid–solid transformation if the liquid is undercooled to a temperature below the equilibrium melting point before solidification begins. The greater the undercooling, the more unstable the undercooled liquid relative to the solid, and hence the greater the tendency for the transformation to occur. From purely thermodynamic reasoning we would expect solidification to begin as soon as the temperature reached the melting point (freezing point). However, a degree of undercooling of the liquid is frequently observed. To understand this phenomenon it is necessary to consider the mechanism of the transformation from liquid to solid. Solidification occurs by the formation of solid nuclei in the liquid. The nuclei are naturally separated from the liquid by a boundary surface, and a certain amount of energy is needed to form this solid/liquid interface. The change in free energy of the system due to the interface is a positive term, $+\Delta G_{interface}$. This energy is opposing the transformation from liquid to solid, which is dependent on the decrease in the bulk or volume free energy of the system, $-\Delta G_{bulk}$. Normally, we can ignore the energy of interfaces in a system since the number of atoms comprising the interface (a grain boundary for example) is very small compared with the number of atoms in bulk. In the case of a solid nucleus formed within the liquid, however, the nucleus is so small that the surface area: volume ratio of the nucleus is relatively large, and we cannot ignore the effect of the interface. For solidification to proceed by growth of the nucleus the negative term, $-\Delta G_{bulk}$, must be greater than the positive term, $+\Delta G_{interface}$. It can be shown* that the condition for the growth of the solid nucleus is that its radius shall be greater than a critical value. If nucleation is homogeneous (*i.e.* it occurs within the system itself) a high degree of undercooling is necessary before solidification can begin. Nickel has been undercooled as much as 319 °C below its freezing point in these circumstances. Since nucleation is usually heterogeneous, in that nuclei for solidification are provided by the crucible containing the molten metal or by insoluble impurities in the melt, undercooling is generally only of the order of a few degrees.

Once stable nuclei are formed, solidification proceeds by the growth of the nuclei at the

---

* J. H. HOLLOMON AND D. TURNBULL, *Progress in Metal Physics*, Vol. IV, Pergamon Press, London, 1953, p. 333; also W. C. WINEGARD, *An Introduction to the Solidification of Metals*, Monograph No. 29, Institute of Metals, London, 1964, p. 9.

expense of the surrounding liquid. The number of grains in the solid metal, and hence its grain size, are directly related to the number of stable nuclei formed during solidification.

The basic aspect of the freezing of alloys in the system illustrated in Fig. 24 is the difference in composition between co-existing liquid and solid phases. The ratio of the solute concentration in the solid phase to that in the liquid phase is called the distribution coefficient, k_0 (Fig. 27a)

$$k_0 = \frac{X_B^s}{X_B^l}.$$

k_0 can vary from just under 1 for very similar components to <0.001 for components which form eutectic systems with very limited solid solubility of the components.

If component B is regarded as the solvent then $k_0 > 1$ (Fig. 27b). Generally, k_0 varies from 1 to 3 in this case.

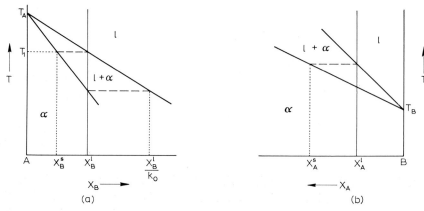

Fig. 27. Part of a phase diagram with (a) $k_0 < 1$, and (b) $k_0 > 1$.

To consider the freezing of an alloy under equilibrium conditions we will start with an alloy whose overall composition is X_B^l (Fig. 28). Freezing will begin when the temperature has dropped to T_1 since at this temperature the liquidus curve is reached. The initial solid separating from the liquid will be α of composition X_B^s. There is a rejection of solute (B) into the liquid phase since the small amount of α formed has a lower solute content than that of the bulk of the liquid from which it was formed. The liquid is enriched with respect to its solute content and the liquid composition must therefore move to the right towards higher solute contents. The liquid now has a lower liquidus temperature than initially and the temperature must fall before further solid can precipitate. For the moment we will assume that the increase in solute content of the liquid is such that the temperature falls to T_2 (Fig. 28) before the liquidus is again reached. At T_2 liquid of composition $l_2 (\equiv X_B^{l2})$ is in equilibrium with solid of composition $s_2 (\equiv X_B^{s2})$. From T_1 to T_2 solid of composition between X_B^s and s_2 will have separated from the liquid. The average composition of the solid will be between X_B^s and s_2. Since l_2 can only co-exist with solid of composition s_2, some solute atoms must diffuse from the liquid into the solid to wipe out the concentration gradient in the solid and move its composition to s_2 at T_2. With the diffusion of solute (B) from the liquid to the solid phase it is apparent that the composition of the liquid phase moves from l_2 to a lower solute content. In order to move the liquid composition back to l_2 an additional quantity of solid of composition s_2 must be precipitated.

Solidification can be treated as a step-wise process throughout the temperature range in which liquid and α co-exist in the same way that has been used for the temperature interval from T_1 to T_2. If each step is considered as an infinitesimally small decrement in temperature the liquidus and solidus appear as continuous curves.

As solidification proceeds, the amount of solid phase increases. According to the lever rule at T_1 the alloy is 100% liquid phase; at T_2 it contains practically 50% solid ($s_2 : l_2 = X_B^{l2} - X_B^l : X_B^l - X_B^{s2}$) and at T_3 the alloy is completely solid. The phase diagram of Fig 28. requires that the final stages in the solidification of alloy X_B^l involve the formation of α of composition $s_3 (\equiv$ the alloy composition, X_B^l) from the last drops of liquid of composition $l_3 (\equiv X_B^{l3})$.

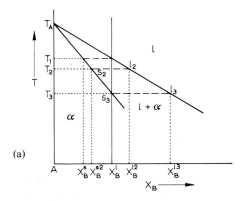

(a)

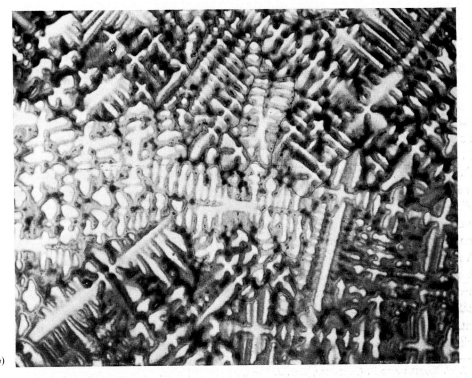

(b)

Fig. 28. Freezing of alloy of composition X_B^l for a phase diagram with $k_0 < 1$. (a) Equilibrium freezing; (b) cored structure in 70/30 brass produced by non-equilibrium freezing, $\times 200$ (courtesy Metallurgical Services).

How can a solid s_3 form from a liquid l_3 which contains so much more component B? The answer depends on three factors: the diffusion of increasing amounts of B from the liquid to the solid which is increasing in quantity at the expense of the liquid as freezing progresses; the reduction in the amount of the liquid phase as freezing proceeds; and the high specific area: volume ratio of the liquid assisting the diffusion process. The diffusion of more and more B from an ever-decreasing amount of liquid has a more pronounced effect on the liquid composition as solidification proceeds. At temperatures near T_3 the amount of liquid transformed to solid by diffusion of B into the α phase becomes greater than that transformed by direct separation. The final stage in solidification can therefore be visualised as the diffusion of a large amount of B into the solid and the separation of a small quantity of solid of composition s_3.

3.2.3. *Non-equilibrium freezing*

The above description of the freezing of an alloy assumes that equilibrium is maintained at all stages. This implies adequate time for diffusion in the α to equalise concentration gradients, and complete mixing in the liquid phase. Under normal conditions of cooling, equilibrium is not maintained. As solidification involves the rejection of solute from the liquid, the liquid adjacent to the first solid to separate will be enriched with respect to component B. Sufficient time is not allowed for the equalisation of composition in the liquid. A concentration gradient is formed in the liquid with a maximum solute content at the solid/liquid interface. Since freezing occurs at the solid/liquid interface, the growth of the initially deposited solid will be from a liquid which is richer in component B than anticipated with reference to Fig. 28. Thus at a temperature T_2 the liquid at the interface will have a composition containing more B than l_2 and this liquid will separate solid richer than s_2. When a temperature T_3 is reached the solid will have an average composition between X_B^s and s_3. Some liquid will remain, and further solidification occurs below T_3 to give a range of solid compositions from s_3 upwards, but the average solid composition on completion of solidification must be equal to s_3 ($\equiv$ overall alloy composition X_B^l). Non-equilibrium freezing therefore occurs over a wider temperature range than equilibrium freezing. There is also a large difference in composition between the first and last solids to separate in non-equilibrium freezing. The resulting concentration gradient in the solid produces what is called a cored structure (Fig. 28b).

The presence of a concentration gradient in the liquid during solidification leads to constitutional undercooling*. It is so named because it occurs when the rate of loss of heat during cooling is such that the temperature in the solute-enriched zone of the liquid is below the equilibrium liquidus temperature. The interpretation of the structure of cast materials in terms of constitutional undercooling is considered by W. C. Winegard**.

Any alloy in this system which has been solidified under non-equilibrium conditions can be restored to equilibrium by elimination of the concentration gradient in the solid phase. This is best achieved by heating to a temperature close to the solidus to allow rapid diffusion and equalisation of concentration. If the cast structure is broken up by working before heat treatment, restoration of equilibrium is facilitated. This is due to the introduction of defects by working which increase the rate of diffusion, reduction of the distance for diffusion by the

---

* J. W. RUTTER AND B. CHALMERS, *Can. J. Phys.*, **31** (1953) 15; W. C. WINEGARD, *Met. Rev.* **6** (1961) 57.
** W. C. WINEGARD, *An Introduction to the Solidification of Metals*, Monograph No. 29, Institute of Metals, London, 1964.

breaking up of the cast structure, and in favourable circumstances to recrystallisation during heat treatment following the initial working.

By comparison with a pure metal it should be noted that alloys in this system (Fig. 24) freeze over a range of temperature. In general, the wider the temperature gap between liquidus and solidus, the greater the tendency for concentration gradients to develop during non-equilibrium freezing (segregation), and the more marked the coring.

3.2.4. *Variants of the simple phase diagram*

The phase diagram of Fig. 24 was derived on the basis of ideal behaviour for both the liquid and solid solutions. We must now consider the effect of departure from ideality. The more general case in alloy systems is that where ΔH_m is positive and $\Delta H_m^{\alpha} > \Delta H_m^{l}$; there is a greater positive departure from ideality for the solid than the liquid solution. Hume-Rothery has linked this with an increasing difference in the atomic radii of the components. This difference introduces a lattice strain energy into the solid solution which is not present in the more open structure of the liquid solution. The strain energy contributes to the enthalpy of mixing term ΔH_m^{α} and results in a tendency to stabilise the liquid rather than the solid phase. Thus the free energy curves for the liquid and solid phases will have the form shown in Fig. 29a at a temperature below the melting points of both components, reflecting the greater stability of the liquid phase.

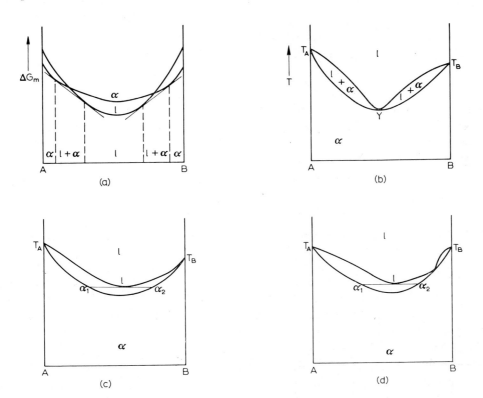

Fig. 29. Variant of Fig. 24 with a minimum in the liquidus. (a) Free energy curves at a temperature $<T_B$; (b) phase diagram; (c) and (d) necessity for a three-phase equilibrium if the compositions of solid and liquid are not identical at the minimum.

A minimum appears in the liquidus and solidus (Fig. 29b) at point Y. At Y the common tangent to the liquidus and solidus is horizontal (*cf.* eqn. (102)). It has already been proved thermodynamically that the composition of two phases in equilibrium in a binary system is identical when the temperature–composition curves for each phase passes through a maximum or a minimum. The same result could have been obtained by application of the Phase Rule. Consider the two alternative possibilities for minima in the $T–X$ curves of two phases. In each case (Fig. 29c and 29d) a tie line can be drawn which allows equilibrium between three phases— a liquid and two solid phases of different composition. The character of the two-phase $l \rightleftharpoons \alpha$ equilibrium is thereby destroyed and we have a three-phase equilibrium. In this instance the invariant reaction $(3+f = 2+1; f = 0)$ is a eutectic. We therefore conclude that, given a two-phase equilibrium, phase diagrams of the type shown in Fig. 29c and d are impossible. Hence, at the minimum the composition of the co-existing phases is identical in agreement with the thermodynamic derivation on p. 44.

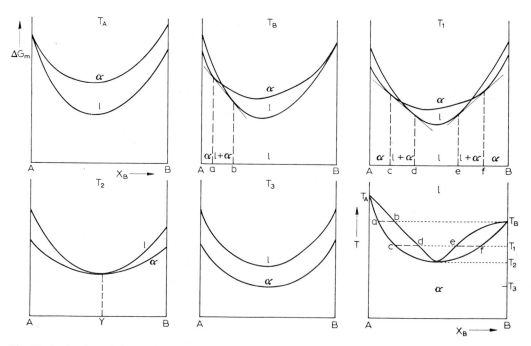

Fig. 30. Derivation of the phase diagram (Fig. 29b) from free energy curves for the liquid and solid phases. $T_A > T_B > T_1 > T_2 > T_3$.

Figure 30 illustrates the sequence of free energy curves leading to a diagram of the type shown in Fig. 29b. The system K–Cs is an example of a system with a minimum in the liquidus. The application of thermodynamics to the calculation of liquid/solid equilibria in binary systems having a minimum in the liquidus is given by Kubaschewski and Heymer[*].

With ΔH_m positive, there is a tendency for phase separation. Thus in many cases there also appears a miscibility gap in the solid state (Fig. 31), as in the Au–Ni system. With increasing ΔH_m the critical temperature increases, since the term C in eqn. (101) increases, leading to a narrower temperature gap between the liquidus minimum and T_c. Between Y and T_c the α solid

* O. KUBASCHEWSKI AND G. HEYMER, *Acta Met.*, **8** (1960) 416.

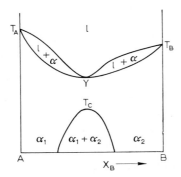

Fig. 31. Phase diagram with a minimum in the liquidus and a miscibility gap in the solid state.

is the stable phase. At temperatures below T_c the α solid solution is less stable than a mixture of phases. These phases have been called α_1 and α_2 in Fig. 31. Tie lines may be drawn in the $\alpha_1 + \alpha_2$ phase field to indicate the phases which co-exist at each temperature from room temperature (the abscissa AB) to T_c. As the temperature increases towards T_c the compositions of the co-existing α_1 and α_2 phases approach one another, until at T_c they are identical. At T_c, $\alpha_1 = \alpha_2 = \alpha$. The difference between α_1 and α_2 is normally one of unit cell parameter; the crystal structures are of the same type. In the Au–Ni system both α_1 and α_2 are face-centred cubic but have different lattice parameters. The designation α, α_1 and α_2 for the single-phase region below the solidus does not therefore imply any reaction between these phases.

It is instructive to apply the Phase Rule to the case of a miscibility gap with a critical phase at temperature T_c. Gibbs* defined a critical phase as follows: "In general we may define a critical phase as one at which the distinction between co-existent phases vanishes". In the $\alpha_1 + \alpha_2$ phase region $f = 1(p+f = c+1; \; 2+f = 2+1)$. Therefore if we fix $X_B^{\alpha_1}$, the Phase Rule informs us that $X_B^{\alpha_2}$ and the temperature for co-existence of the two phases is automatically fixed. At the critical point we place a limitation on the system by requiring the compositions of the co-existing phases to be identical. In this way we have reduced the variables by one, and the system at the critical point will therefore be invariant. This means that the critical point occurs at one temperature and composition only. It is a fixed point.

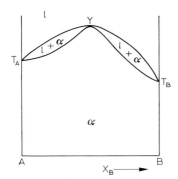

Fig. 32. Phase diagram with a maximum in the liquidus.

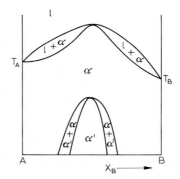

Fig. 33. Appearance of an ordered α' phase at low temperatures.

---

* *The Scientific Papers of J. Willard Gibbs*, Vol. 1, Dover Publications Inc., 1961, (republication of 1st edn. by Longmans, Green & Co., 1906) p. 129.

The alloy whose composition corresponds to Y in Fig. 29b and 31 is said to melt congruently. A congruently-melting alloy is defined as one which melts or freezes at a fixed temperature with no change in the composition of the solid or liquid phase. In this respect a congruent alloy behaves like a pure metal.

If ΔH_m is negative, a maximum occurs in the liquidus and solidus (Fig. 32). With $\Delta H_m < 0$ there is a greater attractive force between dissimilar atoms than between similar atoms, and with $\Delta H_m^\alpha < \Delta H_m^l < 0$ the attractive force would be greater for the solid solution than the liquid. There would be stronger bonds formed in the solid solution, leading to increased melting point of the solid solution compared to the components. At low temperatures the solid will attempt to order itself so as to produce the maximum number of dissimilar atom pairs (Fig. 33). This behaviour follows from the tendency to association of dissimilar atoms rather than similar atoms when $\Delta H_m < 0$. The ordering of the solid solution gives rise to superlattice formation.

3.2.5. *Superlattice formation : order–disorder transformations*

The solid solutions discussed so far have always been characterised by a completely random arrangement of the A and B atoms on the lattice. Under certain conditions (see p. 91) the A and B atoms assume a regular ordered arrangement and the solution is then called an ordered solution or a superlattice. Any change from an ordered to a disordered arrangement of A and B atoms is called an order–disorder transformation.

All superlattices possess one feature in common: A atoms surround themselves with B atoms and vice versa. Figure 34 shows the body-centred cubic β brass superlattice as composed of two inter-penetrating cubic lattices. As will be seen, each Cu atom is surrounded by eight Zn atoms and each Zn atom is similarly surrounded by eight Cu atoms. At high temperatures the entropy effect predominates and a random arrangement of Cu and Zn atoms is produced. This corresponds to the α region in Fig. 33. At low temperatures the energy effect outweighs the entropy effect and the solution is ordered—corresponding to the α' region in Fig. 33*.

It was long believed that order–disorder transformations were not classical phase changes. It was considered that such transformations did not depend on diffusion processes and that

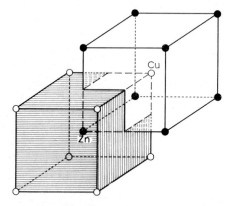

Fig. 34. The β brass superlattice viewed as two inter-penetrating cubic lattices. The atom marked Zn is at the centre of the cube formed by Cu atoms; the atom marked Cu is at the centre of the cube formed by Zn atoms.

---

* For a consideration of the variation of the degree of ordering with temperature in an AB lattice of the β brass (CuZn) type the reader is referred to J. D. Fast, *Entropy*, Cleaver-Hume Press, London, 1962, p. 136.

change of temperature allowed a continuous re-arrangement of the atoms without changing the phase. An ordered region in a phase diagram was therefore shown as a dotted line separating the ordered lattice from the disordered lattice. The Phase Rule could not be applied to the boundary since the Phase Rule requires a two-phase region to be present between the ordered and disordered phases. The work of Rhines and Newkirk* has shown that there are cases in which an ordered phase of one composition exists in equilibrium with a disordered phase of a different composition. There exists a two-phase region between the ordered and disordered phase regions. It is now generally recognised that order–disorder transformations can give rise to classical phase changes which comply with Gibbs' Phase Rule (Fig. 33). The phase boundaries rise to a rounded maximum which usually corresponds to a simple composition of the type AB or AB_3. Only at such simple atom ratios can the transformation (*i.e.* at the temperature maximum) be considered diffusionless.

3.2.6. *Pressure–temperature–composition phase diagram for a system with continuous series of solutions*

The T–X phase diagram (Fig. 24) can be used with the P–T phase diagrams of pure A and B to construct a three-dimensional pressure–temperature–composition (P–T–X) phase diagram as shown in Fig. 35. This appears complicated at first glance but it is only constructed from building blocks with which we are already familiar.

The three axes are P, T and X_B. The back face is the P–T unary phase diagram for component A with an invariant point at O_A; the front face is the P–T unary phase diagram for component B with an invariant point at O_B. The top face is the T–X phase diagram for the binary system A–B. Points T_A and T_B represent the melting points of A and B respectively at a fixed pressure. If we assume that this pressure is atmospheric, T_A and T_B are the normal melting points. The front face is a P–X phase diagram. Such diagrams are little used in alloy studies since, in the main, we are interested in the effects which variation in pressure have on T–X diagrams and not in the effect of variations in temperature on P–X diagrams.

The rest of the lines in Fig. 35 belong to the binary P–T–X equilibrium. The curve $K_A K_B$ joins the critical point of A to the critical point of B. Such critical points as K_A represent the pressure and temperature conditions under which the liquid and vapour phases are indistinguishable. To clarify the liquid $\rightleftharpoons$ vapour equilibrium in the binary phase diagram, conditions at a temperature and pressure below K_B are indicated in Fig. 35 by dotted lines. The three curves joining O_A to O_B and marked (V), (l), and (α) show the effect of a second component on the invariant equilibrium in a unary system. At O_A the liquid, α and vapour phases co-exist and application of the Phase Rule shows that $f = 0$. Addition of component B changes the equilibrium to a monovariant one between l_{AB}, α_{AB}, and V_{AB} phases. Tie lines connect the compositions of liquid, α and vapour phases which co-exist at a particular temperature and pressure. The tie lines are both isothermal and isobaric. The three curves $O_A(V)O_B$, $O_A(l)O_B$ and $O_A(\alpha)O_B$ lie in one plane, and this plane is defined by a series of isothermal–isobaric tie lines. The plane is a three-phase region. This three-phase region is formed by the intersection of three lens-shaped two-phase regions. The latter are the $(l+\alpha)$, $(V+l)$ and $(V+\alpha)$ phase regions. The $(l+\alpha)$ phase region descends from $T_A(l)T_B$ to $O_A(l)O_B$ and from $T_A(\alpha)T_B$ to $O_A(\alpha)O_B$; the $(V+\alpha)$ region from $P_A(\alpha)P_B$ and $P_A(V)P_B$ to $O_A(\alpha)O_B$ and $O_A(V)O_B$ respectively, and the

---

* F. N. RHINES AND J. B. NEWKIRK, *Trans. Am. Soc. Metals*, **45** (1953) 1029.

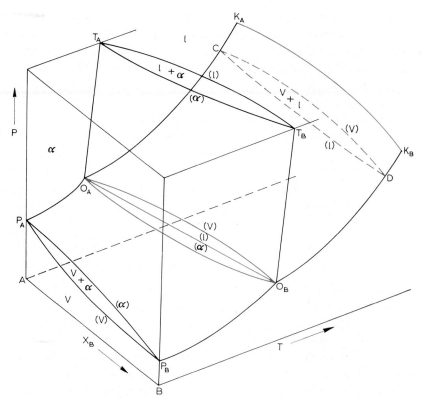

Fig. 35. Pressure–temperature–composition phase diagram for a system with continuous series of solutions

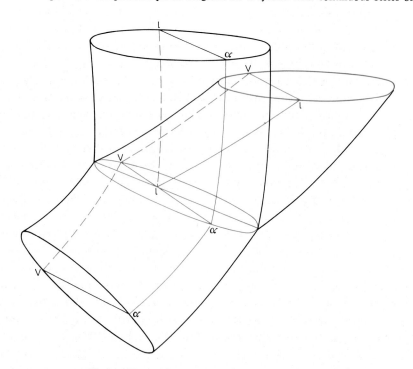

Fig. 36. Formation of a three-phase tie line *Vlα*.

$(V+l)$ region from $C(V)D$ and $C(l)D$ to $O_A(V)O_B$ and $O_A(l)O_B$ respectively. The formation of a tie line in the three-phase region is illustrated in Fig. 36. The dotted lines marked VV and ll lie on the under surface of the $(V+\alpha)$ and $(V+l)$ phase regions and on the back surface of the $(l+\alpha)$ phase region respectively. The full lines marked $\alpha\alpha$ and ll lie on the corresponding top and front surfaces of the phase regions.

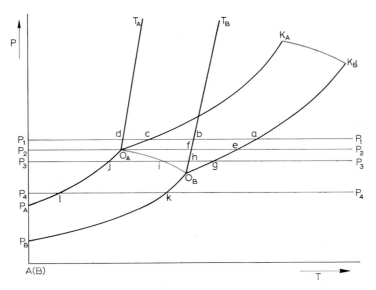

Fig. 37. Two-dimensional projection of Fig. 35 on the $P\text{–}T$ plane for component A.
$P_A O_A$ — equilibrium between V_A and α_A; $P_B O_B$ — V_B and α_B; $O_A T_A$ — l_A and α_A; $O_B T_B$ — l_B and α_B; $O_A K_A$ — V_A and l_A; $O_B K_B$ — V_B and l_B; $O_A O_B$ — V_{AB}, l_{AB} and α_{AB}; O_A — V_A, l_A and α_A; O_B — V_B, l_B and α_B; $K_A K_B$ — $V_{AB} = l_{AB}$.

The equilibria can be presented in a two-dimensional manner by projecting on to the $P\text{–}T$ plane for component A (the back face in Fig. 35). The projection is shown in Fig. 37. As will be seen, the monovariant curves and the invariant point for each unary system project simply. The three-phase region which exists as a plane in Fig. 35 projects as a curve in Fig. 37 since each tie line between V, l and α on the three-phase plane projects as a point on the curve $O_A O_B$. This follows from the isothermal–isobaric nature of the $V+l+\alpha$ tie lines.

To study the effect of a variation in pressure on the form of the $T\text{–}X$ phase diagram we will select four pressure conditions:

$$K_B > P_1 > O_A$$
$$P_2 = O_A$$
$$O_A > P_3 > O_B$$
$$O_B > P_4 > P_B \qquad \text{(see Fig. 37).}$$

T–X diagram at P_1. A horizontal section across Fig. 35 at pressure P_1 will intersect the lens-shaped two-phase region of $V+l$ at high temperatures and the $l+\alpha$ phase region at lower temperatures. The resultant $T\text{–}X$ diagram is shown in Fig. 38. Any alloy in the system consists of the vapour phase at a high enough temperature. As the temperature is lowered liquid condenses from the vapour in exactly the same way as solid is precipitated from the liquid at lower temperatures. The boiling points of pure A and pure B are given by the intersection of the horizontal section with the curves $O_A K_A$ and $O_B K_B$ respectively (*i.e.* as points c and a in

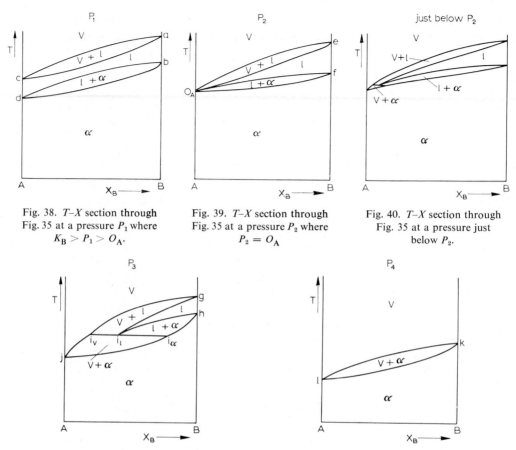

Fig. 38. T–X section through Fig. 35 at a pressure P_1 where $K_B > P_1 > O_A$.

Fig. 39. T–X section through Fig. 35 at a pressure P_2 where $P_2 = O_A$

Fig. 40. T–X section through Fig. 35 at a pressure just below P_2.

Fig. 41. T–X section through Fig. 35 at a pressure P_3 where $O_A > P_3 > O_B$.

Fig. 42. T–X section through Fig. 35 at a pressure P_4 where $O_B > P_4 > P_B$.

Fig. 37). The melting points of pure A and pure B are similarly given by points d and b in Fig. 37. The phase diagram at pressure P_1 is therefore identical with that discussed previously in Fig. 24, the only additional feature being the extension of the temperature axis to include vaporisation processes.

T–X diagram at P_2. An intriguing situation arises if a section is taken at the pressure corresponding to the invariant point, O_A, of component A. The T–X diagram is shown in Fig. 39. Points c and d, representing the boiling point and melting point of A (Fig. 38), now coincide. The $(V+l)$ and $(l+\alpha)$ phase regions have a common origin at point O_A. This type of phase diagram is not common* and we shall in fact notice that generally only two lines issue from a transition point of a pure component. In this particular case, four lines issue from O_A. This diagram can be regarded as the limiting case for the T–X diagram at lower pressures (P_3) insofar as the point O_A can be considered as a degenerate tie line. Point O_A has zero dimensions; the tie line representing equilibrium between V_{AB}, l_{AB} and α_{AB} is one-dimensional. When point O_A is referred to as a degenerate tie line it means that O_A is thought of as a line of zero dimensions. Comparison of Figs. 39, 40 and 41 should clarify this point.

---

* This type of phase diagram has been considered by A. N. Krestovnikov and V. N. Vigdorovich, *Zhur. Fiz. Khim.*, **31** (1957) 1345.

T–X diagram at P_3. A section at P_3 intersects the three-phase region $V_{AB}+l_{AB}+\alpha_{AB}$. The resultant *T–X* phase diagram is shown in Fig. 41. The horizontal line $i_V i_l i_\alpha$ is the isothermal–isobaric tie line at P_3. Such a tie line could be that marked $Vl\alpha$ in Fig. 36. The section at P_3 intersects the three lens-shaped regions $(V+l)$, $(l+\alpha)$ and $(V+\alpha)$ to produce the phase regions $g i_V i_l$, $h i_l i_\alpha$ and $j i_V i_l i_\alpha$ respectively. The disposition of the single-phase fields, V, l and α, in the *T–X* phase diagram can also be deduced from Fig. 35. The point i on curve $O_A O_B$ (Fig. 37) is the projection of the tie line $i_V i_l i_\alpha$ (Fig. 41) on the P–T plane.

The alloys between i_V and i_α undergo a monovariant reaction of the type $l \rightleftharpoons V+\alpha$, the liquid being of composition i_l, the vapour of composition i_V and the solid of composition i_α. This type of reaction involves three-phase equilibrium, which is dealt with in the next chapter.

T–X diagram at P_4. A section at P_4 only intersects the $(V+\alpha)$ lens-shaped phase region. Consequently, the *T–X* phase diagram has the form given in Fig. 42. All alloys condense direct from the vapour phase to the solid on cooling.

CHAPTER 4

Binary Phase Diagrams.
Three-Phase Equilibrium Involving Limited Solubility of the Components in the Solid State but Complete Solubility in the Liquid State

4.1. STRUCTURAL FACTORS

It was stated on p. 43 that complete solubility of two metals in each other is to be expected when these metals have very similar atomic radii, the same valency and crystal structure. The factors which govern the extent of solubility of one metal in another have emerged from the work of Hume-Rothery and the Oxford school*. They can be summarised as follows.

(1) *The atomic size-factor*

If a B atom is substituted for an A atom on the regular A lattice one would expect a distortion of the lattice and internal strain to be generated, if the B atom has a different size from the A atom it replaces. It is reasonable to assume that a relationship will exist between the atomic diameters of the two components and the solubility of B in the A lattice. As postulated by Hume-Rothery, Mabbott and Channel-Evans**, primary solid solutions are restricted to a few atomic percent if the atomic diameters of two metals differ by more than about 14%. The size-factor is said to be unfavourable. This rule is a negative one insofar as the inverse is not necessarily true. A favourable size-factor does not invariably imply extensive solid solubility. The reasons are to be found in the differing electrochemical properties of the metals.

(2) *The electronegative valency effect*

Another factor which restricts solid solubility is the tendency of the two metals to form stable intermediate phases***. The more electropositive one component and the more electronegative the other, the greater the tendency to form an intermediate phase. In the systems of Mg with Si, Sn and Pb, intermediate phases of composition Mg_2Si, Mg_2Sn and Mg_2Pb are formed. The stability of the phases decreases in the order $Mg_2Si > Mg_2Sn > Mg_2Pb$, in agreement with the combination of decreasingly electronegative elements with the electropositive Mg. The solubility of Si, Sn and Pb in Mg is very restricted (<0.1% Si, 3.45% Sn and 7.75% Pb respectively).

The effect of increasing stability of an intermediate phase in lowering the range of α solid solution is illustrated in Fig. 70.

---

* W. HUME-ROTHERY AND G. V. RAYNOR, *The Structure of Metals and Alloys*, Monograph No. 1, 4th edn., Institute of Metals, London, 1962; W. HUME-ROTHERY, *Elements of Structural Metallurgy*, Monograph No. 26, Institute of Metals, London, 1961.
** W. HUME-ROTHERY, G. W. MABBOTT AND K. M. CHANNEL-EVANS, *Phil. Trans. Roy. Soc.*, (*London*), A **233** (1934) 1.
*** For a definition of the term *intermediate phase* see p. 85.

(3) *Relative valency effect*

When the uni-valent metals Cu, Ag and Au are alloyed with metals of higher valency the uni-valent metal has a lower solubility in the metal of higher valency than vice versa. In the Cu–Zn system for example Cu has a maximum solubility of 2.7% in Zn, but Zn is soluble to a maximum of 39% in Cu.

4.2. THREE-PHASE EQUILIBRIUM—EUTECTIC REACTIONS

4.2.1. *The eutectic reaction*

In considering variants of the phase diagram with continuous series of liquid and solid solutions (Fig. 24) it was noted that a positive departure from ideality changed the phase diagram from Fig. 43a to 43b. Greater departure has the effect shown in Fig. 43c and 43d. As mentioned previously (p. 41), most binary systems with a positive value of ΔH_m do not show a miscibility gap with a critical point (Fig. 43b) since melting occurs before the critical temperature is reached, (Fig. 43d). If the components have the same crystal structure, it is permissible to speak of a hypothetical critical temperature indicated by the dotted continuation of the solubility curves in Fig. 43d. Free energy curves for the liquid and solid solutions which lead to a phase diagram of this type are given in Fig. 44.

Comparing Fig. 43b and 43d it is evident that there is no continuous passage from α_1 to α_2; in the latter case there is a two-phase region between the α_1 phase and the α_2 phase. The eutectic system is said to show partial miscibility in the solid state. This type of diagram can also form if the components have different crystal structures. One cannot then speak of a hypothetical critical temperature, and the free energy curves must take into account the different crystal structures of the components. There will be distinct free energy curves for each phase—the liquid, the α and the β phases. The derivation of the phase diagram is given in Fig. 45.

On an atomic basis the sequence of phase diagrams represented by Fig. 43a–d can be viewed as resulting from an increasingly less favourable size-factor. Mixing of the A and B atoms in the solid state to form a solid solution becomes more difficult than mixing in the liquid.

At temperature T_3 (Fig. 45) the liquid, α and β phases are in equilibrium with each other. This temperature is called the eutectic temperature, and the liquid composition at this temperature is referred to as the eutectic composition. The Phase Rule indicates that the co-existence of three phases in a binary system involves an invariant equilibrium. The three phases can

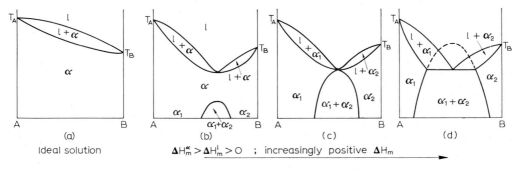

Fig. 43. Effect of increasingly positive departure from ideality in changing the phase diagram for a continuous series of solutions to a eutectic-type.

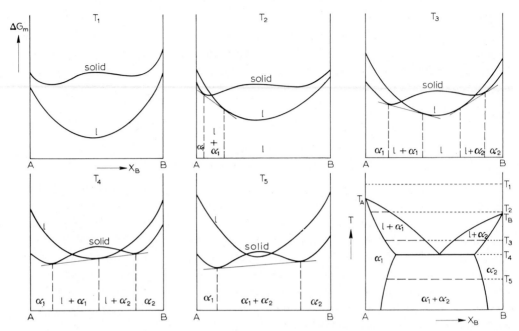

Fig. 44. Derivation of the eutectic phase diagram from the free energy curves for the liquid and solid phases. (After A. H. Cottrell; courtesy Edward Arnold.)

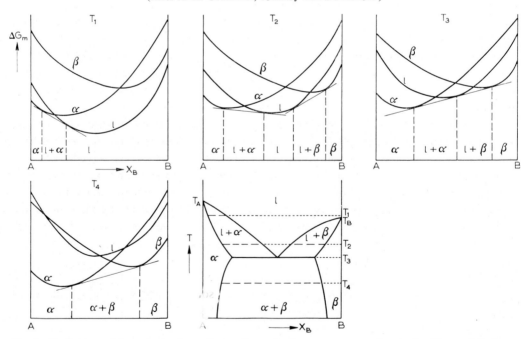

Fig. 45. Derivation of the eutectic phase diagram from the free energy curves for the liquid, α and β phases.

exist at only one temperature, and their compositions are fixed. Therefore an invariant equilibrium in a binary system is represented by a horizontal tie line; the compositions of two phases coincide with the extremities of the tie line and the third phase has an intermediate composition (a, E and b in Fig. 46).

At temperatures below the eutectic temperature, $T_4 < T_3$, the liquid phase has disappeared and the equilibrium is a monovariant one between α and β phases. Therefore in cooling through the eutectic line a reaction of the following type has occurred:

$$l \rightarrow \alpha + \beta.$$

Since the reverse reaction would take place on heating through the eutectic line the reaction is written as a reversible one:

$$l \underset{\text{heating}}{\overset{\text{cooling}}{\rightleftharpoons}} \alpha + \beta.$$

This reaction is the eutectic reaction. Referring to Fig. 46 it simply means that on cooling any alloy whose composition lies between a and b the liquid of composition E simultaneously deposits solid α of composition a and solid β of composition b until all the liquid has been consumed.

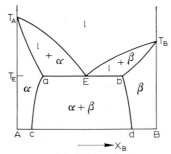

Fig. 46. Binary eutectic phase diagram.

The liquidus curve for the α solid solution is curve $T_A E$ and the solidus curve for the α solid solution is curve $T_A a$; for the β solid solution the liquidus is curve $T_B E$ and the solidus curve $T_B b$. The liquidus in a binary eutectic system is therefore the curve $T_A E T_B$; the solidus is curve $T_A a E b T_B$ which includes the eutectic horizontal aEb. Curves ac and bd are solubility curves (sometimes called solvus curves). The curve ac represents the limit of solid solubility of component B in component A as a function of temperature; curve ac is therefore the solid solubility curve for the α solid solution. Similarly, curve bd is the solubility curve for the β solid solution. At room temperature the lattice of component A can take into solid solution an amount of B equal to c, i.e. $X_B^\alpha = c$; component B can take into solid solution an amount of A which brings the β phase to a composition d, i.e. $X_B^\beta = d$. The two phases are in equilibrium. The term *solubility* therefore refers to the composition of a phase when it is in equilibrium with another phase, but the phase compositions are quoted in terms of the amount of a component (*i.e.* B). It will be noted that the solubility curves slope inwards as the temperature rises, in accordance with the general observation that solubility increases with temperature.

Phase regions

It is customary to call the terminal solid solution on the left-hand side of a binary phase diagram (*i.e.* the solid solution based on component A) the α solid solution. The subsequent phases along the diagram are denoted by successive letters of the Greek alphabet: β, γ, δ, etc.*. In the simple eutectic system there are only two solid phases, α and β.

---

* α alpha, β beta, γ gamma, δ delta, ε epsilon, ζ zeta, η eta, θ theta, ι iota, $\varkappa$ kappa, λ lambda, μ mu, ν nu, ξ xi, o omicron, π pi, ϱ rho, σ sigma, τ tau, υ upsilon, φ phi, χ chi, ψ psi, ω omega.

The α solid solution exists in the area $AT_A acA$. Any point within this area will consist of α. Similarly, the β solid solution exists in the area $BT_B bdB$. The liquid phase exists above the liquidus $T_A ET_B$. There are three two-phase regions. In the case of a two-phase region any point within the region consists of two phases, but the compositions of the phases are represented by points on the boundary lines of this region. Taking the $l+\alpha$ phase region as an example, any point within the area $T_A EaT_A$ is two-phase; the compositions of the phases are given by points on the liquidus $T_A E$ and solidus $T_A a$. In this sense the two-phase regions are empty— only the phase boundaries of such regions are significant.

4.2.2. *Equilibrium freezing of alloys*

The eutectic alloy

The simplest case to consider is an alloy of eutectic composition, E. When cooled from the liquid state (Fig. 46), the alloy begins to freeze at the eutectic temperature, T_E. At this temperature the eutectic reaction occurs:

$$l_E \rightarrow \alpha_a + \beta_b$$

the subscripts referring to the compositions of the phases. Liquid of composition E can be considered as lying on two distinct liquidus curves, $T_A E$ and $T_B E$, and it is necessary for this liquid to deposit both α and β solid phases on cooling. The eutectic reaction proceeds at T_E until all the liquid has been consumed. At this point α_a and β_b will be in equilibrium with each other. The temperature now falls and the compositions of the α and β phases in equilibrium change along the solubility curves ac and bd respectively. The α and β compositions can be obtained at any temperature between T_E and room temperature by drawing a horizontal line in the region marked $\alpha+\beta$ to intersect the solubility curves; the points of intersection give the compositions of the co-existing α and β phases (*cf.* Fig. 45d and 45e). The lever rule can be applied to give the relative amounts of α and β phases at each temperature below T_E.

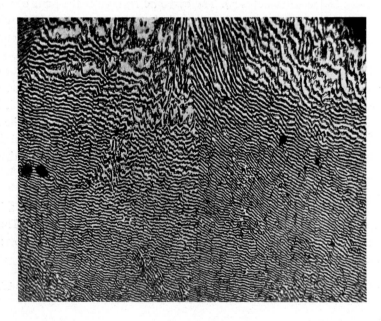

Fig. 47 (a)

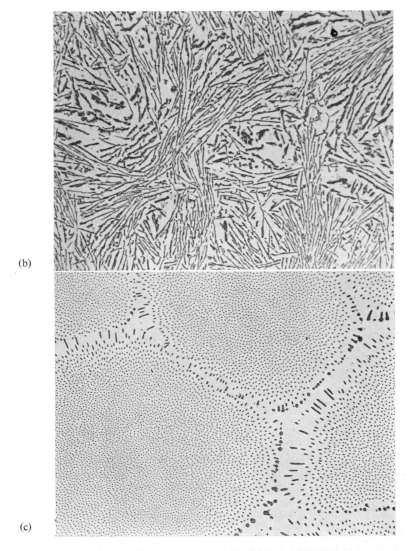

(b)

(c)

Fig. 47. Eutectic structures. (a) Lamellar Pb–Sn eutectic, ×200 (courtesy Tin Research Institute); (b) acicular Al–Si eutectic, ×100 (courtesy I. E. COTTINGTON); (c) globular Cu–Cu$_2$O eutectic, ×200 (courtesy I. E. COTTINGTON).

A eutectic mixture of α and β usually assumes a characteristic microstructure. The structures are called lamellar, globular, rod and acicular (Fig. 47) depending on the appearance of the two phases that constitute the eutectic. This classification of eutectic structures is traditional but somewhat misleading since recent work on the mechanism of eutectic solidification has shown that the structure of a eutectic is dependent upon the conditions of solidification and the effect of impurities in the liquid*. The eutectic liquid begins to solidify by nucleation of one

---

* For a detailed discussion of eutectic solidification the reader is referred to papers by E. SCHEIL, *Z. Metallk.*, **36** (1946) 1; **45** (1954) 298; to W. A. TILLER, *Liquid Metals and Solidification*, Am. Soc. Metals, Cleveland, Ohio, 1958, p. 276; to W. C. WINEGARD, *An Introduction to the Solidification of Metals*, Monograph No. 29. Institute of Metals, London, 1964, p. 65; and to G. A. CHADWICK, Eutectic alloy solidification, *Progr. Mater. Sci.*, **12** (2) (1963).

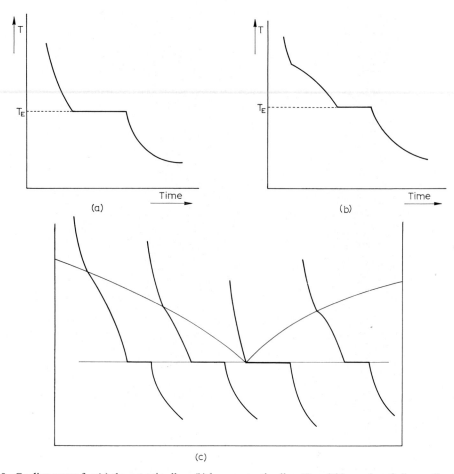

Fig. 48. Cooling curve for (a) the eutectic alloy, (b) hypo-eutectic alloy N, and (c) a series of alloys, allowing the determination of the liquidus and eutectic horizontal.

of the solid phases. If this is the α phase, solute B will be rejected at the interface between the growing α phase and the eutectic liquid. At some stage sufficient enrichment of the liquid at the interface with B atoms will cause the β phase to be nucleated. In lamellar eutectics the β phase nucleates on the α phase. Both phases then grow side by side in the form of lamellae.

If the temperature of the eutectic alloy is plotted as a function of time during cooling from the liquid state a cooling curve of the form shown in Fig. 48a is obtained. This has the same form as the cooling curve for a pure metal.

Hypo-eutectic alloys

A hypo-eutectic alloy is one whose composition, X_B, is less than the eutectic composition. It is conventional to accept the component on the left-hand side of a binary diagram as the reference point for the nomenclature of alloys respective to the eutectic alloy. Thus alloys between component A and the eutectic composition E are called hypo-eutectic alloys. Alloys with compositions between E and component B are called hypereutectic alloys, since they coincide with compositions greater than the eutectic composition.

Consider the freezing of a hypo-eutectic alloy. Three cases can be distinguished. Alloy L

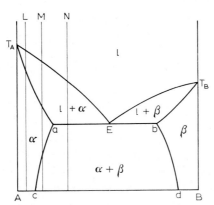

Fig. 49. Freezing of hypo-eutectic alloys.

(Fig. 49) freezes in a manner identical to that discussed for alloys which are soluble in all proportions in the solid state (Fig. 24). At room temperature, alloy L consists wholly of α solid solution. Alloy M (Figs. 49 and 50) freezes as α solid solution, but at lower temperatures a solid state reaction occurs whereby β phase is deposited from the α solid solution. The reactions are clarified in Fig. 50. When solidification begins, solid of composition α_1 separates from liquid of composition l_1. The tie lines $\alpha_2 l_2$, $\alpha_3 l_3$, $\alpha_4 l_4$, indicate the course of solidification which ends when the last drop of liquid, l_4, deposits α of composition α_4. The alloy is now solid α. At a lower temperature it crosses the solubility curve ac. At this temperature α_5 is in equilibrium with β_1, and therefore β_1 is deposited from the α_5 solid solution. With further fall in temperature more β phase is precipitated from the α solid solution, as can be verified by applying the lever rule to the isotherms $\alpha_5\beta_1$, $\alpha_6\beta_2$ and $\alpha_7\beta_3$. Alloy M has a two-phase structure at room temperature. The β phase can precipitate at the α grain boundaries, on certain crystallographic planes in the α phase, or around inclusions and dislocations. The usual microstructure associated with an alloy of composition M is one of grains of α in which the

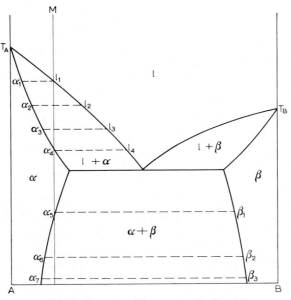

Fig. 50. Freezing of hypo-eutectic alloy M.

β phase has been precipitated along crystallographic planes in the α lattice. Such a structure is called a Widmanstätten structure after Alois de Widmanstätten who first observed this structure in meteorites in 1808*.

In the freezing of alloy M the last drop of liquid has a composition l_4; the liquid does not manage to reach the eutectic composition. Therefore no eutectic reaction occurs during the solidification of alloy M. In fact alloys within the composition range from A to a do not undergo the eutectic reaction when cooled under equilibrium conditions. This is because the liquid is all consumed before the eutectic composition (or temperature) is reached. The limiting case is that for an alloy with composition a. During freezing this alloy deposits α solid solution until the liquid attains the eutectic composition, E. However, at this temperature consideration of the tie line aE shows that only the last drop of liquid manages to reach the eutectic composition. It is probably reasonable to speak of an incipient eutectic reaction rather than to consider that an infinitely small quantity of liquid undergoes the eutectic reaction. At T_E, α of composition a is in equilibrium with β of composition b, i.e. an infinitely small quantity of β. As cooling continues β phase is precipitated from the α matrix in a similar manner to alloy M.

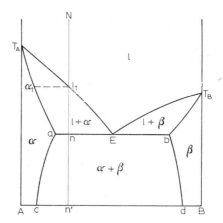

Fig. 51. Freezing of hypo-eutectic alloy N.

Hypo-eutectic alloys between a and E undergo the eutectic reaction. Alloy N (Figs. 49 and 51) begins solidification by depositing α. The liquid changes in composition along the liquidus from l_1 to E, whilst the deposited solid changes its composition from α_1 to a. At T_E there is a proportion an of liquid with eutectic composition and a proportion nE of α, composition a. It is this amount of liquid which now undergoes the eutectic reaction, simultaneously depositing α_a and β_b until the liquid is consumed. When all the liquid has disappeared, the ratio of α to β is nb to an. The α is present as primary α (i.e. that deposited prior to the eutectic reaction) and as one of the phases in the eutectic mixture. The β phase is present only in the eutectic mixture. The metallographic structure of alloy N will therefore be primary α crystals between which there is the characteristic eutectic structure (Fig. 52).

Below T_E alloy N forms more β phase by diffusion of B atoms from the α phase to the eutectic β. That this is so can be verified by considering the change in the ratio of α to β with fall in

---

* More recent information on the orientation relationship between a precipitated phase and the matrix can be found in C. S. BARRETT, *Structure of Metals*, McGraw-Hill, New York, 1952. A very readable account of the Widmanstätten structure is contained in the book by N. T. BELAIEW, *Crystallisation of Metals*, University of London Press, London, 1922.

temperature from T_E to room temperature. At T_E there is an amount *an* of β and *nb* of α; at room temperature the amount of β has increased, as is evident from the ratio of β to α of *cn'* to *n'd*. Not only does more β phase form from the eutectic α; it also forms from the primary α since at T_E both the primary α and the eutectic α have a composition indicated by point *a*. On cooling to room temperature any α of composition *a* will decrease in quantity relative to the β phase. The amount of β phase does not always increase on cooling from T_E to room temperature. If the eutectic composition were relatively close to point *b*, hypo-eutectic alloys

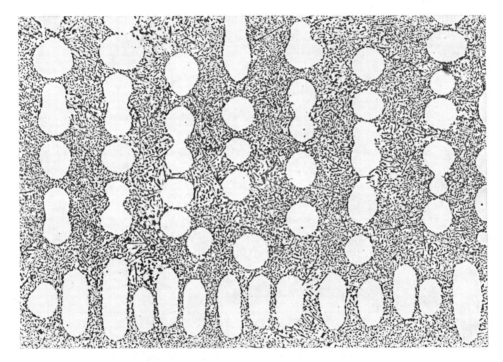

Fig. 52. Structure of a hypo-eutectic alloy. Primary α Al dendrites with an inter-dendritic modified Al–Si eutectic, $\times 300$ (courtesy I. E. Cottington).

with compositions near to the eutectic composition would show an increase in the amount of α on cooling in the solid state. The cooling curve for alloy *N* is given in Fig. 48b. If a series of alloy compositions is taken and cooling curves plotted (the method of thermal analysis), the liquidus and eutectic horizontal can be plotted from the cooling curves (Fig. 48c). It is of interest to note that the first experimental investigation of alloy constitution was Rudberg's work* on the thermal analysis of Pb–Sn alloys.

Hyper-eutectic alloys

Hyper-eutectic alloys behave in a similar way to hypo-eutectic alloys on freezing. Naturally the primary phase to separate is β, followed by the eutectic reaction for alloys with compositions between *E* and *b*.

---

* F. RUDBERG, *Kgl. Svenska Velénskapsakad. Handl.*, **17** (1829) 157, and *Pogg. Ann.*, **18** (1830) 240; commentary by A. PRINCE, *Ann. Sci.*, **11** (1955) 58. The Russian, A. YA. KUPFER (*Ann. Chim. Phys.*, **40** (1829) 285) published a short extract of similar work at the same time as RUDBERG.

Summarising,

alloys between A and c solidify as α solid solution,

alloys between B and d solidify as β solid solution,

alloys between c and a solidify as α but β is precipitated at lower temperatures,

alloys between d and b solidify as β but α is precipitated at lower temperatures,

alloys between a and E deposit primary α followed by a eutectic mixture of $\alpha + \beta$,

alloys between E and b deposit primary β followed by a eutectic mixture of $\alpha + \beta$,

alloy E solidifies isothermally to give a eutectic mixture,

only alloys between a and b undergo the eutectic reaction.

Many metals form simple eutectic systems. Mention may be made of the Ag–Cu, Pb–Sn, Cd–Zn, Pb–Sb and Al–Si binary systems.

4.2.3. *Non-equilibrium freezing of alloys*

Under equilibrium conditions of cooling alloys with compositions between A and a do not undergo the eutectic reaction since the liquid never reaches the eutectic composition. In the more usual condition of non-equilibrium freezing the liquid will often reach the eutectic point E (Fig. 46), although the alloy composition is to the left of point a. Referring to the discussion of non-equilibrium freezing of a solid solution, p. 54, the eutectic temperature will be reached when the liquid has reached point E and the solid is in the form of a cored solid solution. The liquid remaining now freezes as a eutectic mixture. Since the quantity of liquid at T_E will generally be small, the eutectic will form in between the spines of the α dendrites and at the boundaries between different dendrites—the grain boundaries.

4.2.4. *Limiting forms of eutectic phase diagrams*

If the eutectic composition moves towards one of the components the diagram of Fig. 43d changes in the manner indicated Fig. 53. The first case is merely a lateral shift of the eutectic composition towards component B (for example towards Al in the Al–Si system). The further shift exemplified in the Al–Sn system brings the eutectic nearly to component B (for example Sn in the Al–Sn system). The slope of the liquidus curve, dT/dX_B, approaches zero around the middle of the diagram. This condition has been related to the tendency for the liquid solution to separate into two liquids[*]. The oft-quoted limiting condition of Fig. 53c is an impossibility. It implies the existence of complete immiscibility in the solid state between components A and B. In terms of the Phase Rule, a diagram of the type shown in Fig. 53c is ruled out, since it implies that pure component A has a melting range from T_E to T_A. It can also be ruled out by a consideration of the entropy of mixing of a solid solution. The entropy of a pure material is zero at absolute zero. When a solid solution is formed by mixing some B atoms with A atoms to form a very dilute solution of B in A the entropy of the system increases (see p. 5). The increase in entropy, the entropy of mixing, is given by eqn. (18)[**]. The introduction of solute B into pure A increases the entropy and, since a criterion of equilibrium is that the entropy in an isolated system tends to a maximum, it follows that the solution of a solute in a metal is the more stable system than the equilibrium of two pure metals with each other. In other words,

---

[*] L. S. DARKEN AND R. W. GURRY, *Physical Chemistry of Metals*, McGraw-Hill, New York, 1953, p. 330.

[**] Strictly speaking, this is the equation for the entropy of mixing at the absolute zero. It ignores the vibrational or thermal entropy produced by oscillation of atoms in the lattice.

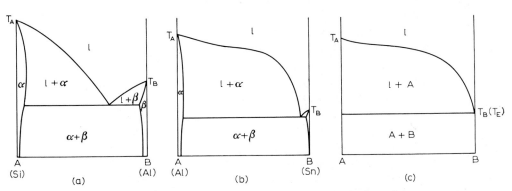

Fig. 53. Evolution of the limiting form of a binary eutectic phase diagram.

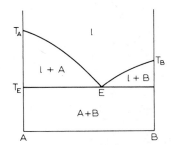

Fig. 54. Impossible form of a binary eutectic phase diagram.

complete immiscibility of two metals does not exist. The solubility of one metal in another may be so low that it is difficult to detect experimentally, but there will always be a measure of solubility. An example occurs in the Ge–Cu system, where the maximum solid solubility of Cu in Ge is less than 10^{-7} atom fraction*. It goes without saying that the eutectic-type diagram shown in Fig. 54 is also an impossibility.

4.2.5. *Retrograde solidus curves*

The normal type of solidus curve slopes inwards from the melting point of the component until it meets an invariant reaction isothermal, such as the eutectic horizontal. The solid solution region increases in width from the melting point to the eutectic horizontal. In 1908 van Laar** predicted the form of solidus curve illustrated in Fig. 55, but it was not until 1926 that the first example of what is called retrograde solubility was found***. A retrograde solidus curve is one which exhibits a maximum solubility of the solute at a temperature between the melting point of the solvent and an invariant reaction isothermal. An alternative definition uses the fact that $dX/dT = 0$ at a temperature T, where $T_E < T < T_A$. There are now something over a dozen systems with definitely established retrograde solubility. Undoubtedly, many others would be found if attention were directed to precise determinations of solidus curves

---

   * C. D. THURMOND AND J. D. STRUTHERS, *J. Phys. Chem.*, **57** (1953) 831.
   ** J. J. VAN LAAR, *Z. Physik. Chem. (Leipzig)*, **63** (1908) 216; **64** (1909) 257.
 *** C. H. M. JENKINS, *J. Inst. Metals*, **36** (1926) 63. Jenkins mistakenly, but understandably, attributed the increasing concentration of Cd in Zn above the eutectic temperature to a peritectic reaction involving an allotropic transformation in Zn and not to retrograde solubility.

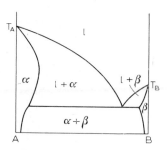

Fig. 55. Retrograde solubility.

in systems with low solubility—at least this has been the outcome in semiconductor research using Ge and Si as solvent metals*.

Thurmond and Struthers have analysed retrograde solidus curves from a thermodynamic standpoint**. They assume an ideal behaviour for the liquid solution and regular behaviour for the solid solution. Expressions for the partial molar free energies of the solute, B, in the liquid and solid solutions are equated. For a dilute solid solution the partial molar free energy of the solute from eqn. (65) is

$$\bar{G}_B^s = (G_B^{\cdot})^s + RT \ln \gamma_B^s X_B^s \tag{108}$$

where $(G_B^{\cdot})^s$ is the molar free energy of pure solute at T.

Since

$$\bar{G}_B^s - (G_B^{\cdot})^s = \Delta \bar{H}_B^s - T \Delta \bar{S}_B^s \tag{109}$$

then,

$$RT \ln \gamma_B^s + RT \ln X_B^s = \Delta \bar{H}_B^s - T \Delta \bar{S}_B^s \tag{110}$$

where $\Delta \bar{H}_B^s$ is the differential heat of solution of the solute ($\Delta \bar{H}_B^s = \bar{H}_B^s - (H_B^{\cdot})^s$) or the enthalpy for transfer of pure B to the dilute solid solution; and $\Delta \bar{S}_B^s = \bar{S}_B^s - (S_B^{\cdot})^s$ is the differential entropy of the solute.

For a dilute solution in which the solute obeys Henry's law $\Delta \bar{H}_B^s$ is independent of composition and temperature and $\Delta \bar{S}_B^s$ is ideal,

$$\Delta \bar{S}_B^s = -R \ln X_B^s. \tag{111}$$

From eqns. (110) and (111),

$$\Delta \bar{H}_B^s = RT \ln \gamma_B^s. \tag{112}$$

Substituting into eqn. (108),

$$\bar{G}_B^s = (G_B^{\cdot})^s + RT \ln X_B^s + \Delta \bar{H}_B^s. \tag{113}$$

Turning to the liquid solution, if this is ideal $\Delta \bar{H}_B^l = 0$, and by analogy with eqn. (113) we can write

$$\bar{G}_B^l = (G_B^{\cdot})^l + RT \ln X_B^l. \tag{114}$$

Equating eqns. (113) and (114), since $\bar{G}_B^s = \bar{G}_B^l$ at temperature T under equilibrium conditions, gives

$$(G_B^{\cdot})^s + RT \ln X_B^s + \Delta \bar{H}_B^s = (G_B^{\cdot})^l + RT \ln X_B^l. \tag{115}$$

---

* C. S. FULLER AND J. D. STRUTHERS, *Phys. Rev.*, **87** (1952) 526 (Cu–Ge); C. D. THURMOND AND J. D. STRUTHERS, *J. Phys. Chem.*, **57** (1953) 831 (Cu–Si and Sb–Ge); R. J. HODGKINSON, *Phil. Mag.*, **46** (1955) 410 (Cu–Ge).

** J. L. MEIJERING (*Philips Res. Rept.*, **3** (1948) 281) has also considered retrograde solidus curves. He approached the problem by a study of the entropy of the liquid and solid solutions.

If ΔH_B is the heat of fusion of pure solute B and ΔS_B is its entropy of fusion, using pure solid as the reference state

$$(G_B^{\cdot})^l - (G_B^{\cdot})^s = \Delta H_B - T\Delta S_B. \tag{116}$$

Substituting in eqn. (115),

$$\ln \frac{X_B^s}{X_B^l} = \frac{\Delta H_B - \Delta \overline{H}_B^s}{RT} - \frac{\Delta S_B}{R} \tag{117}$$

Both X_B^s and X_B^l vary with temperature, but our primary concern is to obtain an expression for X_B^s—the concentration of solute in the dilute solid solution of B in A. To take care of X_B^l in eqn. (117) we must consider the behaviour of the solvent A. From eqn. (104),

$$\ln X_A^s - \ln X_A^l = \frac{\Delta H_A}{R}\left(\frac{1}{T} - \frac{1}{T_A}\right).$$

As X_A^s in a dilute solution is very nearly equal to one,

$$-\ln X_A^l = -\ln (1 - X_B^l) = \frac{\Delta H_A}{R}\left(\frac{1}{T} - \frac{1}{T_A}\right)$$

or

$$\ln (1 - X_B^l) = \frac{\Delta H_A}{R}\left(\frac{1}{T_A} - \frac{1}{T}\right). \tag{118}$$

The value of X_B^l obtained from eqn. (118) may now be substituted in eqn. (117) to allow a calculation of X_B^s as a function of temperature. Naturally we also require values for ΔH_A, ΔH_B, $\Delta \overline{H}_B^s$, and ΔS_B. The entropy of fusion of metals does not vary a great deal and ΔS_B can be given a value of 3 entropy units. Since the heats of fusion of metals are known, this leaves $\Delta \overline{H}_B^s$ to be evaluated.

In most alloy systems the value of $\Delta \overline{H}_B^s$ is relatively small. The higher $\Delta \overline{H}_B^s$, the greater the

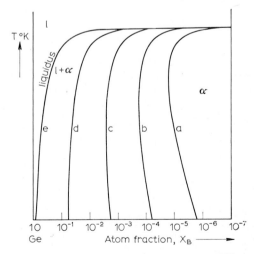

Fig. 56. Solidus curves calculated from eqns. (117) and (118) assuming $\Delta \overline{H}_B^s = 22{,}000$ cal/g.-atom (curve a), 16,500 cal/g.-atom (curve b), 11,000 cal/g.-atom (curve c), 5,500 (curve d). Curve e is the liquidus calculated from eqn. (118). (After C. D. THURMOND AND J. D. STRUTHERS, *J. Phys. Chem.*, **57** (1953) 832; courtesy American Chemical Society.)

tendency to retrograde solubility. This fact can be demonstrated by calculating the solidus curve for various assumed values of $\Delta \overline{H}_B^s$. Taking solvent A with $\Delta H_A = 8,100$ cal/g.-atom, and $T_A = 1210\,°\mathrm{K}$, and a solute B with $\Delta H_B = 3,500$ cal/g.-atom, the liquidus curve is readily calculated from eqn. (118). The solidus curve for B in solution in A is calculated from eqns. (117) and (118) for assumed values of $\Delta \overline{H}_B^s$ equal to 5,500, 11,000, 16,500 and 22,000 cal/g.-atom. The results are given in Fig. 56*. It can be seen that high values of $\Delta \overline{H}_B^s$ yield retrograde solidus curves. If there is a large difference between the melting points of A and B, the heat of fusion of the solute could be sufficiently small for it to be ignored when applying eqn. (117). In this case, with $T_A \gg T_B$ and $\Delta H_A \gg \Delta H_B$, retrograde solidus curves could appear at lower values of $\Delta \overline{H}_B^s$ than previously.

From the thermodynamic viewpoint we conclude that retrograde solidus curves appear when $\Delta \overline{H}_B^s$ has a high value. In these circumstances it is not necessary for there to be a large difference in the melting points of the components. If there is such a difference, retrograde solidus curves can appear even though $\Delta \overline{H}_B^s$ is not unduly high. A high value of $\Delta \overline{H}_B^s$ is associated with a significant difference in atomic radii for A and B, which can lead to a large strain energy contribution to the heat of solution. The difference in atomic radii leads to very low solubilities of the solute in the solvent, in agreement with the general observation that retrograde solidus curves are associated with low solubilities.

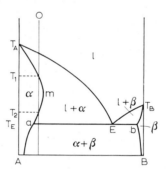

Fig. 57. Partial re-melting associated with retrograde solubility.

The phenomenon of partial re-melting occurs during the solidification of alloys whose composition lies between the solubility limit at the eutectic temperature, point a, and the maximum solubility at a higher temperature, point m, (Fig. 57). Consideration of alloy O shows that it is completely solid at T_1 and remains so until it again crosses the solidus curve at T_2. At T_2 it partially re-melts, the amount of liquid formed increasing until the eutectic horizontal is reached. At T_E the small quantity of liquid undergoes the eutectic reaction and the alloy once more becomes completely solid.

4.2.6. *Disposition of phase boundaries at the eutectic horizontal***

Examination of the phase diagrams for eutectic systems given previously will show that if the solidus curve and the solubility curves are extended, the extensions are into two-phase regions

* C. D. THURMOND AND J. D. STRUTHERS, *J. Phys. Chem.*, **57** (1953) 832.
** The treatment of H. LIPSON AND A. J. C. WILSON (*J. Iron Steel Inst.*, **142** (1940) 107) is followed in this section.

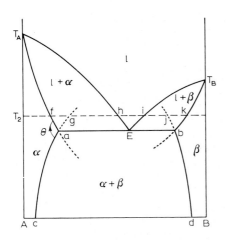

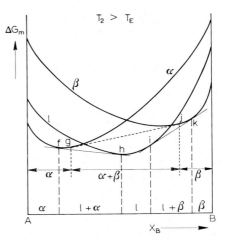

Fig. 58. Disposition of phase boundaries at the eutectic horizontal aEb.

Fig. 59. Free energy curves for the liquid, α and β phases at a temperature T_2 where $T_2 > T_E$.

(Fig. 58). The solidus curve extends into the $\alpha + \beta$ phase region; the solubility curve into the $l + \alpha$ phase region. To demonstrate the validity of this statement we may consider the relative dispositions of the free energy curves for the liquid, α and β phases at T_E and just above T_E. The free energy curves are given in Fig. 45 ($T_E = T_3$ and $T_2 > T_E$). Figure 59 reproduces the curves for T_2 with the addition of a dotted line to represent the common tangent at T_2 to the free energy curves for α and β. The α phase boundary corresponds to point f; the β phase boundary to point k. If, however, the liquid phase were not present at T_2, and only α and β were in equilibrium, the α composition would be given by point g and the β by point j. Points g and j are corresponding points on the extensions of the solubility curves for the α and β phases respectively (*cf.* Fig. 58 with Fig. 43d). Since point g lies to the right of point f the solubility curve ca must lie to the right of the solidus curve $T_A a$. Similarly, by considering the free energy curves at a temperature below T_E it can be shown that the extension of the solidus curve $T_A a$ lies to the right of the solubility curve ca. Therefore the angle θ between the solidus and solubility curves must be less than 180°. The extensions of the phase boundaries at the eutectic horizontal lie in two-phase regions. This is a general rule applicable to all curves which meet at an invariant reaction horizontal in a binary diagram, whether they be eutectic, peritectic, eutectoid, etc., horizontals. The construction illustrated in Fig. 60 is impossible.

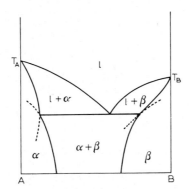

Fig. 60. Impossible dispositions of phase boundaries at a eutectic horizontal.

4.3. THREE-PHASE EQUILIBRIUM—PERITECTIC REACTIONS

4.3.1. *The peritectic reaction*

If there is a considerable difference between the melting points of the components and a positive departure from ideality such that $\Delta H_m^\alpha > \Delta H_m^l > 0$, the phase diagram with a continuous solid solution can change in the way illustrated in Fig. 61. The Au–Pt system has a phase diagram of type Fig. 61b. With greater departure from ideality the peritectic type of phase diagram is produced (Fig. 61d). Above the peritectic horizontal the liquid phase is in equilibrium with α_1; below liquid is in equilibrium with α_2. The derivation of Fig. 61d from free energy curves is illustrated in Fig. 62*.

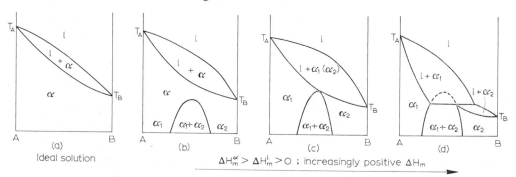

Fig. 61. Effect of increasingly positive departure from ideality in changing the phase diagram from a continuous series of solutions to a peritectic-type.

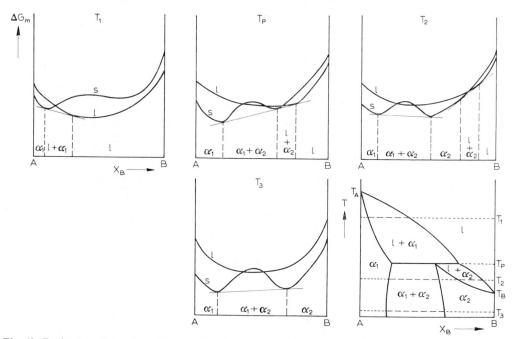

Fig. 62. Derivation of the peritectic phase diagram from the free energy curves for the liquid and solid phases.

---

* Where the two solid phases have different crystal structures there will appear free energy curves for liquid, α and β phases. Reference may be made to A. H. COTTRELL (*Theoretical Structural Metallurgy*, Edward Arnold, London, 1955) for a derivation of the phase diagram.

At T_1, liquid exists in equilibrium with α_1; fall in temperature to T_P brings liquid, α_1 and α_2 into equilibrium with each other. T_P is the temperature of the invariant peritectic reaction. At T_2, α_1 is in equilibrium with α_2 and liquid is in equilibrium with α_2. The invariant peritectic reaction is therefore a reaction between liquid and α_1 to give α_2:

$$l + \alpha_1 \rightarrow \alpha_2.$$

That this is so can be confirmed by considering the reaction during heating. The α_2 phase decomposes into liquid and α_1. Thus the peritectic reaction is written as

$$l + \alpha_1 \underset{\text{heating}}{\overset{\text{cooling}}{\rightleftharpoons}} \alpha_2$$

or, with solid phases α and β,

$$l + \alpha \rightleftharpoons \beta.$$

It will be noticed that there is a simple relation between the eutectic and peritectic reaction insofar as the arrangement of the phase boundaries is concerned. The peritectic is the image of the eutectic (Fig. 63).

Fig. 63. Relationship between eutectic and peritectic reactions.

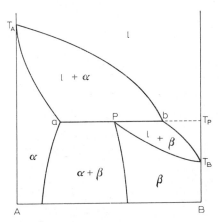

Fig. 64. Binary peritectic phase diagram.

Referring to Fig. 64, the peritectic horizontal is the isotherm aPb. P is called the peritectic point; it corresponds to the composition of the β phase formed by the peritectic reaction in contrast to a eutectic point, which corresponds to the composition of liquid from which two solid phases are precipitated. The line $T_A b T_B$ is the liquidus and the line $T_A a P T_B$ the solidus. The system Ag–Pt is of the simple peritectic type.

4.3.2. Equilibrium freezing of alloys

As in a eutectic system, three types of alloy can be distinguished. These are the alloy with the peritectic composition, P, and those alloys with concentrations of B less than P ($X_B < P$) and greater than P ($X_B > P$). By analogy with eutectic phase diagrams the alloys can be referred to as peritectic, hypo-peritectic and hyper-peritectic respectively, although this designation is not normally used. The terms hypo- and hyper- are usually reserved for differentiating alloy compositions in systems containing eutectic and eutectoid reactions.

The peritectic alloy

The peritectic alloy (Fig. 65) meets the liquidus curve at point l_1 when the temperature of the melt has fallen to T_1. Solid α of composition a_1 separates and thereby increases the solute content of the liquid. Freezing proceeds normally with an increasing amount of α separating as the liquid composition moves along the liquidus from l_1 to b and the α changes in composition along the solidus from a_1 to a. At T_P the peritectic alloy therefore consists of a fraction Pb/ab of α and a fraction aP/ab of liquid. The peritectic reaction occurs at T_P between the liquid remaining (aP/ab) and the primary α deposited from T_1 to T_P (Pb/ab):

$$l_b + \alpha_a \rightarrow \beta_P.$$

On completion of the reaction, all the liquid and all the α have been consumed and the alloy is 100% β phase. That this is so can be confirmed by examining the changes which occur when the peritectic alloy is heated from room temperature to T_P. At room temperature the alloy consists of a fraction ed/cd of α and a fraction ce/cd of β. The amount of α decreases on heating and it is readily seen that, at T_P, the α phase has disappeared and the alloy is wholly β. If heating were continued, naturally the β would undergo the peritectic reaction to produce l_b and α_a.

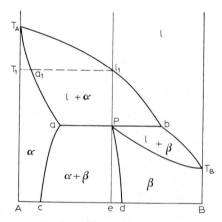

Fig. 65. Freezing of the peritectic alloy P.

On cooling the peritectic alloy from T_P to room temperature α is precipitated from the β. At room temperature the alloy consists of a fraction ed/cd of α and ce/cd of β. If the phase diagram Fig. 65 had been drawn with the solubility curve Pd sloping to the left of Pe the peritectic alloy would react to form 100% β and would remain unchanged with fall in temperature.

Alloys with $X_B < P$

Alloys with compositions between A and c freeze as α solid solutions. Alloys with compositions between c and a freeze initially as α and then precipitate a small amount of β from the α when the solubility curve ca is crossed. Both types of alloy behave similarly to those in the eutectic system (cf. p. 71).

Alloys with compositions between a and P undergo the peritectic reaction, since these are the only alloys, for the condition $X_B < P$, where the liquid reaches point b before it is totally consumed. Alloy L (Fig. 66) begins to solidify when liquid l_1 deposits an infinitely small amount of α of composition a_1. Subsequently, the liquid moves along the liquidus curve from l_1 to b, whilst the solid moves from a_1 to a along the solidus curve. At T_P, just before the peritectic reaction begins, there is a much smaller amount of liquid than solid (ratio of liquid/solid $=$ af/fb). All the liquid reacts with some of the solid α to yield β. The alloy, after the peritectic reaction, is composed of the residue of the primary α plus β formed peritectically. The ratios of α to β are fP to af. In alloys between a and P there is an excess of α remaining after the

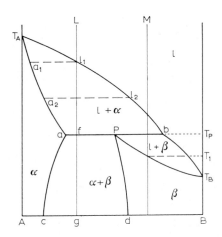

Fig. 66. Freezing of alloys L and M in the peritectic system.

peritectic reaction. This is obvious from a consideration of the phase regions through which the alloy L passes on cooling from the liquid state. Above T_P it is in the $l+\alpha$ phase region; below T_P it is in the $\alpha+\beta$ phase region. Alternatively, remembering that $l/\alpha = aP/Pb$ for the peritectic alloy, we can write for alloy L,

$$l/\alpha = af/fb = (aP-fP)/(Pb+fP),$$

i.e. there is an excess of α at T_P compared to that required for a complete consumption of liquid and α for the peritectic alloy.

From T_P to room temperature alloy L consists of α and β. The relative amounts of α and β change with fall in temperature as with alloy N in the eutectic system (Fig. 51) p. 73.

Alloys with $X_B > P$

For alloys with compositions from b to B the β phase is formed directly from the liquid. Alloys between P and b undergo the peritectic reaction. In such alloys there is an excess of liquid phase over that required to react with the primary α. After the peritectic reaction therefore an alloy such as M (Fig. 66) consists of $l+\beta$ of composition b and P respectively. With further cooling the liquid changes in composition along bT_B and the β along PT_B until, at T_1, the last drop of liquid solidifies and the alloy is wholly β phase. All alloys between P and b solidify in the same way as alloy M, but those with compositions between P and d precipitate α from the β phase when they cross the solubility curve Pd.

4.3.3. *Non-equilibrium freezing of alloys*

The effect of departure from equilibrium conditions during cooling can be considered by using the peritectic alloy as a specific example. On cooling alloy *P* under equilibrium conditions the liquid and α should react to produce a completely β phase structure. When reaction begins between the liquid and α, the β is produced at the interface between the liquid and α phases (Fig. 67a). This β phase acts as a barrier to further reaction between liquid and α. Completion

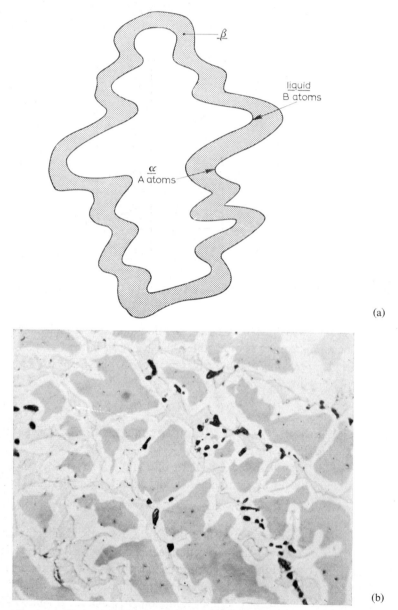

(a)

(b)

Fig. 67. (a) The peritectic reaction, and (b) arc-melted Th–33 at. %Ge alloy. The primary α phase (Th$_3$Ge$_2$—dark grey) has undergone reaction with the matrix (light grey) to form a rim of β phase (The$_3$Ge—white). The black particles are ThO$_2$ impurity. (After A. Brown and J. J. Norreys, *J. Less-Common Metals*, **5** (1963) 302; courtesy Elsevier Publishing Co.)

of the reaction can only occur by allowing sufficient time for the diffusion of A atoms from the α to the β phase and of B atoms from the liquid to the β phase. Such diffusion will allow growth of β at the α/β and β/liquid interfaces until the alloy is wholly β. If equilibrium conditions are not maintained, the primary α phase will not be completely converted to β. The primary α will persist although it is not an equilibrium phase. As shown in Fig. 67b, a reaction rim of β phase surrounds the non-equilibrium α phase*.

4.3.4. *Formation of intermediate phases by peritectic reaction*

A common type of phase diagram is one containing an intermediate phase which is formed by peritectic reaction between the liquid and a terminal solid solution (Fig. 68). An example is

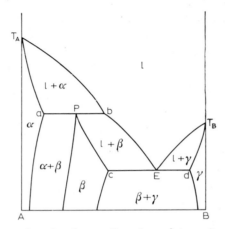

Fig. 68. Formation of an intermediate phase, β, by peritectic reaction.

to be found in the Hf–W system with the formation of an intermediate phase based on the composition HfW_2 by peritectic reaction**. In this case, the intermediate phase is formed by the reaction:

$$l_b + \alpha_a \rightleftharpoons \beta_P \qquad \text{(Fig. 68)}$$

or,

$$l + \alpha \rightleftharpoons HfW_2$$

where α is the terminal solid solution, and β is the intermediate phase. There is also a eutectic reaction:

$$l_E \rightleftharpoons \beta_c + \gamma_d$$

where γ is the terminal solid solution based on component B.

An intermediate phase, as its name implies, has a composition intermediate between that of the terminal solid solutions. It generally has a crystal structure which is different from those of the components. In older literature intermediate phases were regarded as of fixed composition like chemical compounds. They were consequently called intermetallic compounds. However, in most cases the compositions of intermediate phases do not conform to what would be expected from valency considerations, nor are the compositions fixed.

---

* In fact, Fig. 67b is an example of a peritectoid reaction.
** H. BRAUN AND E. RUDY, *Z. Metallk.*, **51** (1960) 360.

Intermediate phases can be broadly classified as either normal valency compounds, size-factor compounds or electron compounds.

(1) Normal valency compounds

Intermediate phases which obey the valency rules are called normal valency compounds. Examples are the phases Mg_2Si, Mg_2Sn, Mg_2Pb and Mg_3Sb_2. They have much in common with ionic compounds such as NaCl and CaF_2, and are sometimes called partly-ionic compounds.

(2) Size-factor compounds

Some intermediate phases have structures which are determined by the relative size of the constituent atoms. The most well-known are the so-called Laves phases*, which are intermediate phases based on the formula AB_2, where atom A has the larger atomic diameter. Laves phases crystallise in structures which are basically arrays of tetrahedra on which the small B atoms are situated and within which there is sufficient space to accomodate the larger A atoms.

In addition, there is a range of size-factor compounds which depend on the insertion of atoms in the interstices of the host lattice. These interstitial compounds are formed when the atomic diameters of the two atom types differ greatly. The metal carbides, nitrides and borides are of this type.

(3) Electron compounds

When the components have similar electrochemical properties and a favourable size-factor, the composition of many intermediate phases occurs at one of three valency electron-to-atom ratios. Such phases are termed electron compounds.

In each of the systems Cu–Zn, Cu–Ga, and Cu–Sn an intermediate phase β is found at a composition based on the formulae CuZn, Cu_3Ga, and Cu_5Sn. The number of valency electrons associated with each element is taken to be

Cu 1; Zn 2; Ga 3; Sn 4

giving a valency electron-to-atom ratio, or electron concentration, for these β phases of 3 : 2. They are known as 3 : 2 electron compounds. Although they occur at different compositions in terms of the atomic %Cu in the appropriate phase diagrams, they all three appear at the same electron concentration and they all have similar crystal structures** (body-centred cubic β brass).

Electron compounds also occur at electron concentrations of 21 : 13 and 7 : 4. The former have a γ brass structure which is a complex cubic lattice with 52 atoms per unit cell. The latter have a close-packed hexagonal structure similar to ε brass.

A definitive treatment of the factors affecting the formation of intermediate phases in alloy systems is given by Hume-Rothery and Raynor***.

Excepting normal valency compounds, in alloy systems we nowadays regard intermediate

---

* F. LAVES, *Theory of Alloy Phases*, Am. Soc. Metals, Cleveland, Ohio, 1956, p. 124; W. HUME-ROTHERY AND G. V. RAYNOR, *The Structure Of Metals and Alloys*, Monograph No. 1, 4th edn., Institute of Metals, London, 1962, p. 228.

** 3 : 2 electron compounds also occur with a complex cubic, or βMn, and a close-packed hexagonal structure.

*** W. HUME-ROTHERY AND G. V. RAYNOR, *The Structure of Metals and Alloys*, Monograph No. 1, 4th edn., Institute of Metals, London, 1962, p. 174.

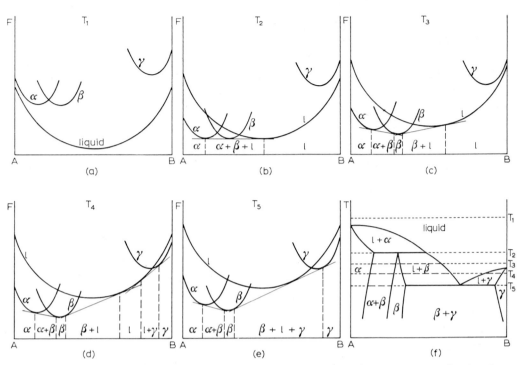

Fig. 69. Derivation of the phase diagram (Fig. 68) from the free energy curves of the liquid, α, β and γ phases. (After A. H. COTTRELL; courtesy Edward Arnold.)

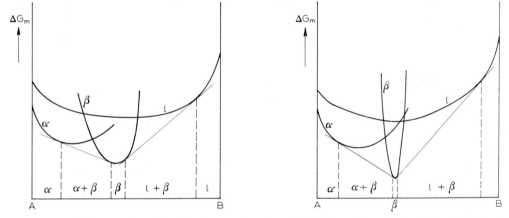

Fig. 70. Decreasing range of stability of an intermediate phase with its increasing stability relative to the terminal solid solutions.

phases as phases akin to terminal solid solutions. They may have an ordered arrangement of atoms in contrast to the terminal solid solutions, or a random atomic arrangement. Their composition may be very limited, or they may show appreciable solubility for the components*.

The intermediate phase in Fig. 68 forms incongruently. By reaction between a liquid of

---

* The Russian metallurgist N. S. KURNAKOV introduced the name "Daltonide" for a phase with a fixed composition and "Berthollide" for a phase with variable composition, but this terminology has remained essentially Russian. Reference should be made to G. S. SMITH, *Metal Ind. (London)*, **68** (1946) 451, 471, 495, 516.

composition b and a solid of composition a the intermediate phase β of composition P is formed. The β phase exists over a range of composition at lower temperatures. The phase diagram can be derived from free energy curves as shown in Fig. 69. To produce this type of phase diagram requires that the free energy curve of the β phase move downwards relative to those of the liquid and α phases as the temperature falls (Fig. 69a–c), and also that the free energy curve of the γ phase moves downwards relative to the liquid curve to establish the eutectic equilibrium (Fig. 69a–d). The form of the free energy curve for intermediate phases shows a rapidly increasing slope at each side. In other words, the composition range for stability of the intermediate phase is rather limited, since there is a rapid increase of free energy outside this composition range, Fig. 70.

A remarkably good example of peritectic reactions occurs in the Cu–Zn system. There are five successive peritectic reactions (Fig. 71):

$$l + \alpha \rightleftharpoons \beta$$
$$l + \beta \rightleftharpoons \gamma$$
$$l + \gamma \leftrightharpoons \delta$$
$$l + \delta \rightleftharpoons \varepsilon$$
$$l + \varepsilon \rightleftharpoons \eta$$

The reaction $l + \beta \rightleftharpoons \gamma$ is of particular interest insofar as it can be considered a limiting case of the peritectic equilibrium in which the peritectic point virtually coincides with the liquid composition (Fig. 72). Thermodynamically, points P and b do not coincide since we have

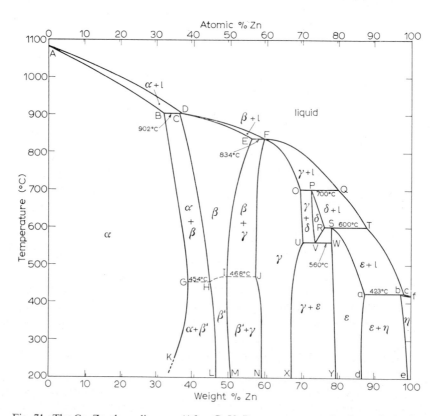

Fig. 71. The Cu–Zn phase diagram. (After G. V. RAYNOR; courtesy Institute of Metals.)

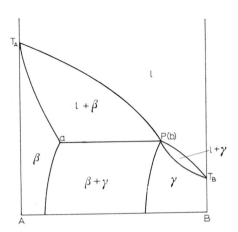

Fig. 72. Limiting case of the peritectic reaction.

already noted the the condition for a liquid and a solid phase to have identical compositions is that a temperature maximum or minimum must be present (p. 44). No such maximum or minimum occurs in the peritectic phase diagram. We must conclude that the peritectic point and the liquid composition are so close to each other that the experimental techniques* used were not able to distinguish them. More refined methods would be expected to produce evidence of a compositional difference between these two points.

Another point worthy of note is the decreasing solubility of Zn in Cu with rise in temperature, in contrast to the normal decrease in solubility with fall in temperature (*cf.* p. 40). A decreasing solubility limit with rise in temperature is sometimes associated with an equilibrium with a disordered intermediate phase (*e.g.* the β phase above 454 °C, Fig. 71). This has been explained as being due to a greater relative movement of the free energy curve of the intermediate phase compared with the α solid solution with rise in temperature (Fig. 73). At a temperature T_2,

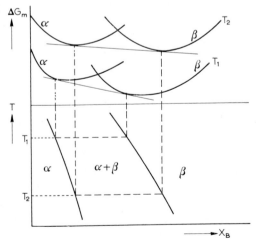

Fig. 73. Illustration of a decrease in solubility with rise in temperature by means of free energy curves.
(After A. H. COTTRELL; courtesy Edward Arnold.)

---

* In compiling the Cu–Zn phase diagram, G. V. RAYNOR (*Institute of Metals Annotated Equilibrium Diagrams*, No. 3, 1944) used the experimental work of R. RUER AND K. KREMERS (*Z. Anorg. Chem.*, **184** (1929) 193) and J. SCHRAMM (*Metallwirtschaft*, **14** (1935) 995, 1047) to delineate this region of the diagram.

the tangent to the free energy curves of α and β gives the α and β equilibrium compositions. With rise in temperature the free energy curves move downwards since the free energy of all phases decreases with rise in temperature. The free energy curve of the intermediate phase β falls more rapidly than that of the α phase, since β is a disordered solution which occurs at a composition nearer the equiatomic ratio ($X_B = 0.5$). The β phase will therefore have a higher entropy of mixing than α, and this fact will be reflected in the more rapid fall in its free energy (since $\Delta G_m = \Delta H_m - T\Delta S_m$). At a temperature $T_1(T_1 > T_2)$ the tangent to the free energy curves of α and β will have a greater negative slope and the equilibrium compositions of α and β will therefore have moved nearer component A.

4.3.5. Non-stoichiometric compounds*

The intermediate phase β in Fig. 68 was stated to have the composition P. The range of existence of the β phase indicates that it is capable of existing with excess A atoms and excess B atoms. In general, intermediate phases appear at some simple atomic ratio of A : B atoms, e.g. A_2B, AB, AB_2, etc. In most cases the phase region for the intermediate phase includes this stoichiometric composition. In the Hf–W system the intermediate phase is HfW_2. The phase region for existence of HfW_2 includes the stoichiometric composition. There are, however, examples of intermediate phases which do not include the stoichiometric composition, although their crystal structure is clearly derived from a stoichiometric compound. For example, the compound FeO (wüstite) exists over a range of compositions in the Fe–O phase diagram, but this composition range does not include FeO. In fact, wüstite contains an excess of O in the lattice. At 1200 °C its composition varies from $FeO_{1.049}$ to $FeO_{1.174}$**. The θ phase in the Al–Cu system is based on the composition $CuAl_2$, but the composition limits of θ lie at higher Al contents*** than indicated by the stoichiometric formula. An alloy with the stoichiometric composition is two-phase. Other examples of non-stoichiometric phases can be found in the literature†.

An explanation for the occurrence of the non-stoichiometric θ phase in the Al–Cu system was given by Rushbrooke‡ and its general relevance to departures from stoichiometry in alloy phases was noted by Lipson and Wilson§. Consider the phase diagram given in Fig. 74. The compound A_2B is in the two-phase region of $\alpha+\beta$ but the β phase, which is based on the stoichiometric compound A_2B, exists at higher B contents. Making the reasonable assumption that the free energy curve of the β phase has a minimum at the stoichiometric composition A_2B we can examine the relative dispositions of the free energy curves for the α, β and γ phases. If these have the configuration shown in Fig. 75, the common tangent to the α and β curves will meet the β curve at a point to the right of the minimum on this curve. The most stable state for the alloy A_2B is then a mixture of α and β; homogeneous β phase occurs at higher B contents. If the free energy curve for α were above that for β, as in Fig. 69, the stable state for alloy A_2B would be as single-phase β. In this case the β phase region includes the stoichiometric composition A_2B.

* An excellent summary of current thought is given by J. S. ANDERSON (*Proc. Chem. Soc.*, (1964) 166–173).
** L. S. DARKEN AND R. W. GURRY, *J. Am. Chem. Soc.*, **68** (1946) 798.
*** D. STOCKDALE, *J. Inst. Metals*, **52** (1933) 111.
† See, for example, R. J. HODGKINSON, *J. Electron.*, **1** (1956) 612; **2** (1956) 201; J. S. ANDERSON, *The Physical Chemistry of Metallic Solutions and Intermetallic Compounds*, (National Physical Lab. Symposium) Paper 7A, H.M.S.O., 1959; D. A. ROBINS, *ibid.*, Paper 7B.
‡ G. S. RUSHBROOKE, *Proc. Phys. Soc. (London)*, **52** (1940) 701.
§ H. LIPSON AND A. J. C. WILSON, *J. Iron Steel Inst.*, **142** (1940) 107 (see p. 111).

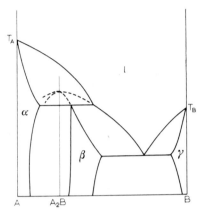

Fig. 74. A non-stoichiometric β phase based on
the intermediate phase A_2B.

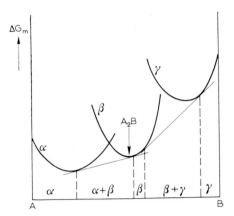

Fig. 75. Use of free energy curves to illustrate the
occurrence of non-stoichiometric phases.

Since only one free energy curve can have its minimum below that of other free energy curves only one phase region can exist which includes the stoichiometric composition of an intermediate phase, irrespective of the number of intermediate phases in the binary system.

4.4. CONGRUENT TRANSFORMATIONS

Three examples of congruent transformation have been noted already: a melting point minimum, a melting point maximum, and a critical temperature associated with a $\beta \rightarrow \alpha$ order–disorder transformation. These are summarised in Fig. 76a and b (*cf.* Figs. 29b and 33). The formation of a superlattice has been related to a negative value for ΔH_m (p. 58)—a greater attractive force between dissimilar atoms than between similar atoms. From the structural viewpoint superlattices are formed when there are significant electrochemical and size-factor differences. A large electrochemical factor expresses the tendency for the solute atoms to avoid each other and to associate with the solvent atoms. A size-factor just within the favourable limit would be expected to lead to atomic re-arrangement so as to relieve the lattice distortion imposed by the solute atoms.

A congruent transformation can also appear with the formation of an intermediate phase (Fig. 76c and d). This can be understood as a consequence of an increasingly negative ΔH_m, with ΔH_m^α more negative than ΔH_m^l (Fig. 77). Alternatively, it can be considered to be a result of an increasing difference in electrochemical properties of the components.

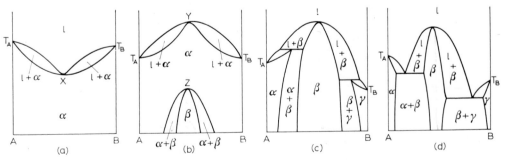

Fig. 76. Examples of congruent transformations.

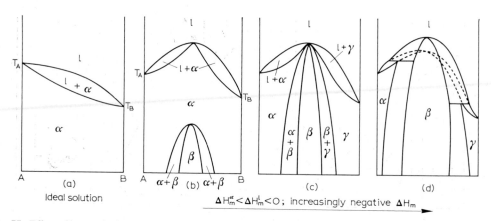

Fig. 77. Effect of increasingly negative departure from ideality in changing the phase diagram from a continuous series of solutions to one containing a congruent intermediate phase.

Under the conditions leading to the phase diagram of Fig. 76c the β phase is sufficiently stable so that its melting point is higher than the maximum in the liquidus (Fig. 76b), a point illustrated in Fig. 77d.

The more usual type of diagram in which a congruently-melting intermediate phase exists is shown in Fig. 76d. In this case the phase diagram can be considered as two simple eutectic systems placed side by side, the congruently melting alloy acting in a sense as a component. Thus in Fig. 78 we can speak of the partial phase diagrams A–X and X–B. In many cases the intermediate phase in a diagram such as that given in Fig. 78 is a normal valency compound. Examples are found in the systems Mg–Si, Mg–Sn, and Mg–Pb with congruent phases (normal valency compounds) Mg_2Si, Mg_2Sn, and Mg_2Pb. Laves phases are also particularly stable compounds and often appear as congruently-melting compounds. The Laves phase UFe_2 can be taken as an example*. Kubaschewski** has used the U–Fe system as an example for the calculation of the phase diagram from the heats and entropies of formation of the various phases. The width of the β phase region in Fig. 78 has been exaggerated for clarity. In the three Mg systems mentioned there is very little solubility of the components in the β phase and the phase diagram is often drawn as a limiting case (Fig. 79).

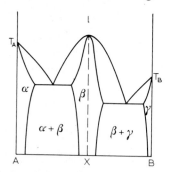 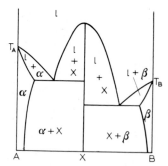

Fig. 78. Phase diagram with a congruent intermediate phase.   Fig. 79. Limiting case of Fig. 78.

---

* J. D. GROGAN AND C. J. B. CLEWS, *J. Inst. Metals*, **77** (1950) 571.

** O. KUBASCHEWSKI, Symposium on *Thermodynamics of Nuclear Materials*, International Atomic Energy Agency, Vienna, 1962, p. 219.

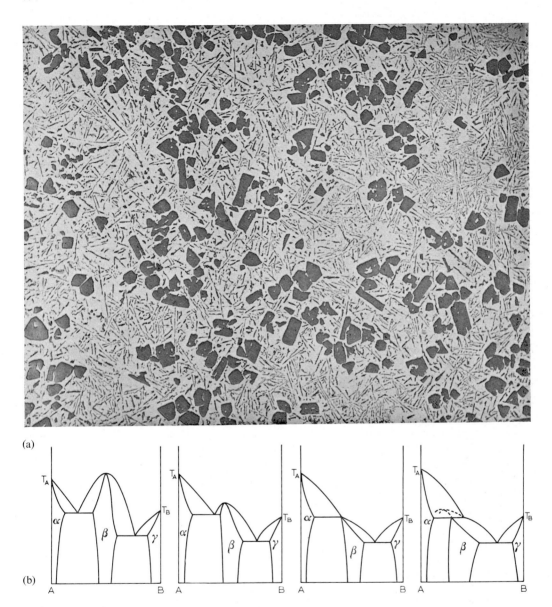

Fig. 80. (a) Microstructure of a cast Al–22 % Si alloy showing polyhedra of primary Si (dark grey phase) in an overmodified eutectic matrix, ×90 (courtesy The British Non-Ferrous Metals Research Association). (b) Relationship between phase diagrams containing congruent and incongruent intermediate phases.

The solidification of alloys in a system of the type shown in Fig. 78 is very similar to that discussed for eutectic alloy systems. The only difference is in the appearance of the primary phase when this is the β phase. Primary β often separates idiomorphically. A well-defined primary dendrite structure does not form, but the primary phase separates with well-formed crystal facets (Fig. 80a).

The relationship between the phase diagrams with congruent and incongruent intermediate phases is shown in Fig. 80b.

CHAPTER 5

Binary Phase Diagrams.
Limited Solubility in Both the Liquid and Solid State

5.1. INTRODUCTORY THERMODYNAMICS

In all the phase diagrams considered so far there has been a complete miscibility in the liquid state, even when there has been very limited solid solubility. In terms of the size-factor this may be considered as being due to the ability of a liquid to accommodate atoms of different atomic diameter to a greater extent than a solid. In going from a condition of complete solubility in the liquid state to partial solubility one must necessarily pass through a state of incipient immiscibility. This process has already been noted for solid solutions. In Fig. 14 free energy–composition curves were plotted for various values of kT/C. At low temperatures, $kT/C < 0.5$, the free energy curves show two points of inflection, and we have noted that such curves correspond to the existence of a solid miscibility gap in which the inflections give the compositions of the co-existing phases. At higher temperatures, $kT/C > 0.5$, there are no inflections and the curve falls continuously from $X_A = 0$ to $X_A = 0.5$ and from $X_A = 1$ to $X_A = 0.5$. There is complete solubility between the components. At $kT/C = 0.5$ there appears

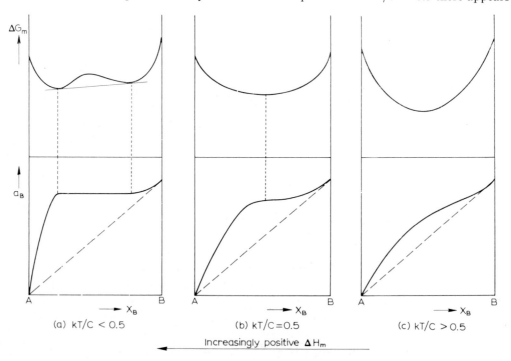

Fig. 81. Free energy and activity curves for (a) $kT/C < 0.5$, (b) $kT/C = 0.5$, and (c) $kT/C > 0.5$.

the critical case representing the limiting condition for the appearance of a miscibility gap (*cf.* eqn. (101)). The discussion of free energy of solutions was based on a consideration of solid solutions, but exactly the same treatment could have been applied to liquid solutions.

The appearance of immiscibility in the liquid state can also be traced by deriving activity curves from the free energy–composition curves, as illustrated in Figs. 19 and 20. The free energy curves and their corresponding activity curves for values of kT/C of <0.5, $= 0.5$ and >0.5 are given in Fig. 81. The horizontal inflection in Fig. 81b indicates the conditions for incipient immiscibility with fall in temperature from $kT/C > 0.5$ to $kT/C < 0.5$. Figure 81 can also be used to indicate the manner in which immiscibility can occur in some binary systems and not in others. Figure 81c corresponds to a system showing positive deviations from ideality (*cf.* Fig. 7c). With increasing ΔH_m the system will show greater deviations as indicated in Fig. 81b, and with very large positive deviations liquid immiscibility results. The sequence of phase diagrams given in Fig. 43 can be considered as being extended to give the sequence shown in Fig. 82. In Fig. 82b there is a horizontal inflection in the liquidus curve; a greater

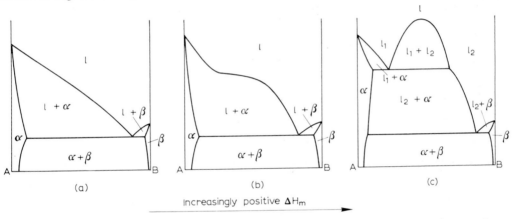

Fig. 82. Effect of very large positive deviations from ideality in changing the phase diagram from a eutectic to a monotectic.

positive departure from ideality leads to liquid immiscibility (Fig. 82c). Such a sequence of phase diagrams is to be found in the systems Au–Tl, Ag–Tl and Cu–Tl, corresponding to Figs. 82a–c respectively.

To refresh our memories, a positive enthalpy or heat of mixing ΔH_m implies:

(1) $H_{AB} > \frac{1}{2}(H_{AA} + H_{BB})$

i.e. H_{AB} is less negative than $\frac{1}{2}(H_{AA} + H_{BB})$.

(2) $A \leftrightarrow B < \frac{1}{2}(A \leftrightarrow A + B \leftrightarrow B)$

i.e. similar atoms attract more strongly than dissimilar atoms.

(3) $\gamma_A > 1$ and $\gamma_B > 1$

i.e. since the activity coefficients are greater than 1 ($\gamma_A = a_A/X_A > 1$) there are positive departures from ideality, the degree of departure reflecting the value of ΔH_m as shown in Fig. 81.

An example of the appearance of a liquid miscibility gap and the activity curves corresponding to temperatures above, at, and below the critical temperature is given by Lumsden* for

---

* J. LUMSDEN, *Thermodynamics of Alloys*, Monograph No. 11, Institute of Metals, London, 1952, p. 340 (cf. Figs. 104 and 108).

the Pb–Zn system. A critical review of the application of thermodynamics to the Pb–Zn system is given by Rosenthal, Mills and Dunkerley*.

Attempts have been made to relate the onset of liquid immiscibility in binary systems to physical properties of the components. Axon** suggested that a wide difference in melting point and atomic diameter of the components should lead to liquid immiscibility. Hildebrand and Scott*** related immiscibility to the heats of vaporisation of the components (a measure of the binding energies) and their atomic volumes. Immiscibility was predicted when

$$\tfrac{1}{2}(V_A + V_B)\left[\left(\frac{\Delta H_A^v}{V_A}\right)^{\frac{1}{2}} - \left(\frac{\Delta H_B^v}{V_B}\right)^{\frac{1}{2}}\right] < 2RT$$

where V_A and V_B are the atomic volumes of A and B, and ΔH_A^v and ΔH_B^v are the heats of vaporisation of A and B respectively. Both these approaches give no more than a guide to the prediction of liquid immiscibility. As Hildebrand and Scott admitted, there are many binary systems which comply with their inequality but which form completely miscible liquids. It was suggested that in such systems the immiscibility tendency was counterbalanced by a tendency towards association of dissimilar atoms. Mott† modified the Hildebrand and Scott inequality to take this opposing factor into account. Although more successful in predicting liquid immiscibility there are still many exceptions to be explained.

5.2. MONOTECTIC REACTIONS

5.2.1. *General*

We are now in a position to consider the phase diagram given in Fig. 82c. The appearance of a miscibility gap in the liquid state leads to a binary invariant reaction of the type

$$l_1 \rightleftharpoons \alpha + l_2.$$

This reaction, in which a liquid decomposes on cooling to yield a solid phase and a new liquid phase, is called a monotectic reaction (Fig. 83). In the latter figure tie lines have been drawn in the two-phase regions to indicate the compositions of the co-existing phases.

Fig. 83. The monotectic reaction.

Above the monotectic temperature, T_M, two two-phase regions occur. The one originating at T_A (Fig. 84) is the $l_1 + \alpha$ phase region; the other is the liquid miscibility gap represented by the $l_1 + l_2$ phase region. From the monotectic temperature to the critical temperature, T_c, the two liquids are immiscible, and will separate into two layers. One layer corresponds in composition to points on the line Mk and the other layer to points on the line bk. Isothermals (tie lines)

   * F. D. ROSENTHAL, G. J. MILLS AND F. J. DUNKERLEY, *Trans. AIME*, **212** (1958) 153.
  ** H. J. AXON, *Nature*, **162** (1948) 997.
*** J. H. HILDEBRAND AND R. L. SCOTT, *Solubility of Non-Electrolytes*, Reinhold, New York, 1950.
   † B. W. MOTT, *Phil. Mag.*, **2** (1957) 259.

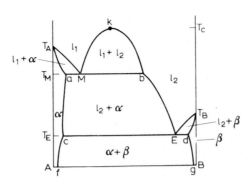

Fig. 84. Binary phase diagram incorporating a monotectic reaction.

connect the compositions of co-existing phases and it is evident that the length of the tie lines, and hence the compositional difference between the two liquid phases, decreases with rise in temperature from the monotectic temperature. At T_c the tie line has degenerated to a point (*cf.* p. 56 for a discussion of solid state miscibility gaps) and the two liquid phases l_1 and l_2 are now indistinguishable. The upper part of the liquid phase region which shows complete miscibility of the components is therefore given the designation *l*.

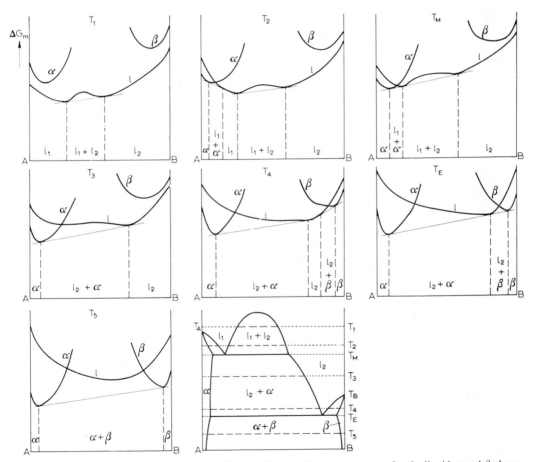

Fig. 85. Derivation of the monotectic phase diagram from the free energy curves for the liquid, α and β phases.

Below the monotectic temperature a single two-phase region exists. This is the $\alpha + l_2$ phase region. To illustrate a complete diagram with consumption of all the liquid l_2, Fig. 84 has been drawn with a eutectic reaction occurring at a lower temperature

$$l_{2(E)} \rightleftharpoons \alpha_c + \beta_d.$$

This is usually the case in practice, as for example in the Ga–Pb and Ga–Tl systems*. The low temperature invariant reaction could be peritectic in nature but this is not usual.

The liquidus is the line $T_A MbE T_B$ and the solidus is the line $T_A ac Ed T_B$. The sequence of free energy curves leading to the phase diagram (Fig. 84) is shown in Fig. 85.

5.2.2. *Equilibrium freezing of alloys*

The monotectic alloy

The monotectic alloy, M in Fig. 84, begins to freeze when the temperature reaches T_M by undergoing the invariant monotectic reaction

$$l_{1(M)} \rightarrow \alpha_a + l_{2(b)}.$$

This reaction is similar to that of the eutectic reaction with the β phase replaced by l_2. Since the monotectic composition is invariably closer to the composition a of the α solid solution than to the composition b of l_2, the α phase predominates in the monotectic mixture. As drawn, the ratio of α to l_2 is about 5 : 1. The α forms the matrix in which the l_2 phase is dispersed. With continued cooling from T_M to T_E more α is deposited from l_2 as the α moves along the solidus ac and l_2 along the liquidus bE. This α will deposit on the monotectic α and will be indistinguishable from it. At T_E the liquid reaches the eutectic composition E and undergoes the invariant eutectic reaction

$$l_E \rightarrow \alpha_c + \beta_d.$$

It should be noted that the eutectic composition is usually very close to the low-melting point component (B) and therefore the eutectic liquid deposits an overwhelming amount of β compared with α. In the Cu–Pb system, for example, the compositions corresponding to points c, E and d are <0.09, 99.80 and $>99.99\%$ Pb respectively, so that the eutectic contains 99.8% β and only 0.2% α. The final structure of the solid monotectic alloy is thus one consisting of a matrix of α with dispersed particles of eutectic. Eutectics with compositions close to one of the components normally do not exhibit a typical structure. They form divorced eutectics, *i.e.* the α in the eutectic is nucleated by the existing α matrix and the β in the eutectic is nucleated at random in the remaining liquid. Divorced eutectics do not present the normal geometrical arrangement of phases typical of normal eutectic structures.

Alloys with $X_B < M$

Alloys with compositions between a and M begin to freeze by depositing primary α in dendritic form. When the liquid reaches the monotectic composition, α of composition a and l_2 of composition b are formed. The monotectic α is deposited on the primary α and the alloy now consists of interlocking dendrites of α with a small amount of liquid l_2 dispersed between these dendrites. With further cooling more α is deposited from the interdendritic liquid l_2

---

* B. PREDEL, *Z. Metallk.*, **50** (1959) 663.

until the temperature falls to T_E. Liquid l_2 then decomposes in a eutectic manner as before to give a divorced eutectic structure.

The final structure of the "hypo-monotectic" alloys is thus one of interdendritic divorced eutectic between an α dendritic matrix.

Alloys with $X_B > M$

Only alloys with compositions between M and b are worthy of consideration. Such alloys on cooling from a sufficiently high temperature will break down into two liquid layers. Depending on the overall alloy composition, one layer will have a composition represented by a point on the miscibility curve Mk and the other layer a composition on the miscibility curve bk. Since the densities of the two liquids usually differ, one layer will form on top of the other. In the case of the Cu–Pb alloys the Cu-rich liquid l_1 forms the top layer and the Pb-rich layer l_2 the bottom layer. If alloys within the miscibility gap are cooled rapidly, there will be insufficient time for separation into layers and an emulsion will be formed, *i.e.* very small droplets of one liquid suspended in the other liquid.

Just above the monotectic temperature slowly cooled alloys will show a top liquid layer of composition approaching M and a bottom liquid layer of composition approaching b. At the monotectic temperature only the top layer of composition M decomposes into α_a and $l_{2(b)}$; the bottom layer remains unchanged as $l_{2(b)}$. On cooling from T_M to T_E more α is precipitated from the liquid l_2. In the top layer this α will separate on to the pre-existing monotectic α; in the bottom layer the α separates from l_2 as discrete particles. At T_E the remaining liquid transforms eutectically. The top layer will have a structure consisting of α and pools of divorced eutectic (formed from liquid $l_{2(E)}$); the bottom layer will consist of small particles of α in a divorced eutectic matrix. The relative amounts of α and divorced eutectic in the top layer is dependent on the overall alloy composition. The nearer this is to M the greater the amount of α.

Microstructural features of Cu–Pb alloys have been discussed by Pelzel*.

5.2.3. *Limiting forms of monotectic phase diagrams*

Some metals, such as Cu and W, are virtually insoluble in each other in both the solid and liquid state, *i.e.* the solubility is so small as to be undetected experimentally to date. Such phase diagrams as the Cu–W diagram can be considered as limiting cases of the monotectic phase diagram (Fig. 86).

In deriving the phase diagram in Fig. 86d it will have been noted that the monotectic point M approaches the composition of pure A and the monotectic temperature approaches T_A. Similarly, the composition of l_2, and hence the eutectic composition E, approaches the composition of pure B and the eutectic temperature approaches T_B. In the Cu–W system no alloys are formed. Referring to Fig. 86d, component A represents W and component B represents Cu. If Cu and W are melted together they form two liquid layers. On cooling, the bottom layer (W) solidifies first at the freezing point of W and the top layer solidifies subsequently at the freezing point of Cu.

* E. PELZEL, *Metall*, **10** (1956) 1023.

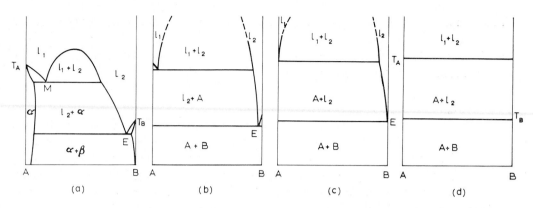

Fig. 86. Limiting form of the monotectic phase diagram.

5.3. SYNTECTIC REACTIONS

The syntectic reaction occurs only rarely in alloy systems. The reaction involves the conversion of two liquid phases into a solid phase on cooling:

$$l_1 + l_2 \rightleftharpoons \alpha.$$

In those few systems in which syntectic reactions occur—K–Zn, Na–Zn, K–Pb, Pb–U, and Ca–Cd—the α phase is an intermediate phase in which the components are hardly soluble at all. An idealised phase diagram is shown in Fig. 87a and, for comparison, the K–Zn phase diagram in Fig. 87b. The solubility ranges for the components and the intermediate phase, KZn_{13}, are so small that they cannot be included in a scaled diagram.

Any alloy with an overall composition in the range a to b (Fig. 87a) will form two liquid layers, l_1 and l_2, on cooling. According to the phase diagram when such alloys cool to the syntectic temperature, T_S, the two liquids should react together to produce solid α

$$l_{1(a)} + l_{2(b)} \rightleftharpoons \alpha_{(S)}.$$

This reaction will proceed at the interface between the two liquid layers. The α formed will act as a barrier to continued reaction between l_1 and l_2. In general, therefore, it will be very

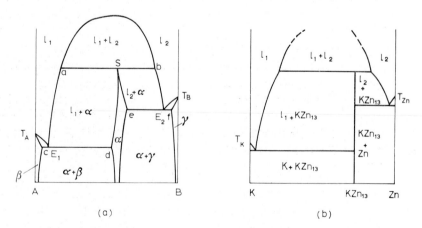

Fig. 87. Syntectic phase diagrams. (a) Schematic; (b) the K–Zn system.

difficult to maintain equilibrium conditions in a syntectic system. After the formation of a quantity of α the two liquid layers will tend to freeze independently of each other. The l_1 layer will deposit more α as the liquid composition moves along the liquidus curve aE_1 and, on reaching E_1, this liquid would react eutectically to form $\alpha + \beta$—again as a divorced eutectic structure since E_1 is so close in composition to component A. The l_2 layer would behave similarly, its composition changing along bE_2 and freezing ending with the eutectic reaction $l_{E_2} \rightleftharpoons \alpha + \gamma$.

The liquidus in this system is the line $T_A E_1 a S b E_2 T_B$ and the solidus is the line $T_A c E_1 d S e E_2 f T_B$.

Binary Phase Diagrams.
Reactions in the Solid State

6.1. SOLID STATE REACTIONS

There are three invariant solid state reactions of immediate interest:

(1) The eutectoid reaction, $\alpha \rightleftharpoons \beta + \gamma$

the solid state analogue of the eutectic reaction (Fig. 88a).

(2) The monotectoid reaction, $\alpha_1 \rightleftharpoons \beta + \alpha_2$

the solid state analogue of the monotectic reaction (Fig. 88b).

(3) The peritectoid reaction, $\alpha + \beta \rightleftharpoons \gamma$

the solid state analogue of the peritectic reaction (Fig. 88c).

The only difference between the three solid state reactions and the analogous reactions in which a liquid phase or phases are present is that the rates of reaction in the solid state transformations re slower than when liquid is present. Only the kinetics of the reactions differ; thermodynamically the eutectoid decomposition of a solid solution, for example, can be treated in exactly the same way as the eutectic decomposition of a liquid phase. In considering liquid → solid transformations on p. 51, mention was made of the need to consider the positive free energy contribution arising from the formation of an interface between the solid nuclei

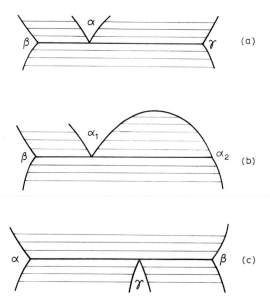

Fig. 88. Solid state reactions. (a) The eutectoid reaction; (b) the monotectoid reaction; (c) the peritectoid reaction.

and the liquid. With a solid → solid transformation, such as the eutectoid reaction which also proceeds by a nucleation and growth mechanism (*cf.* eutectic solidification on p. 70), an additional factor has to be taken into account. This is the strain energy, $+\Delta G_{strain}$, which is induced in the material as a result of dimensional changes occurring during transformation. The change in dimensions is a consequence of differing crystal structures and lattice parameters for the precipitating phase compared with the matrix phase. Since the matrix is rigid the dimensional changes will induce elastic strain in the matrix. It can be seen that the strain energy factor opposes the transformation. The transformation can only proceed if $-\Delta G_{bulk} > +\Delta G_{interface} + +\Delta G_{strain}$.

Since grain boundaries are regions of high free energy they are prone to be first to become unstable during transformation. The disordered atomic arrangement at grain boundaries will reduce the strain energy factor and the interfacial energy needed to nucleate a new phase. Nucleation is therefore frequently observed to begin at the grain boundaries. The finer the grain size, and hence the larger the grain boundary area, the more readily will the transformation proceed.

Many solid state invariant reactions occur when one or both components undergo allotropic transformations, although allotropy of the components is not a necessary condition for the appearance of solid state reactions.

6.2. EUTECTOID REACTIONS

Examples of eutectoid reactions are found in the Fe–C and Al–Cu binary systems (Fig. 89). In the former case the eutectoid reaction involves two phases based on allotropic modifications of Fe, whereas in the latter case the eutectoid reaction is not associated with an allotropic transformation of a component.

In view of the technological importance of the eutectoid reaction in the Fe–C system this phase diagram will be considered in some detail. Pure Fe exists in three forms—α, γ and δ. From room temperature to 910 °C the body-centred cubic α Fe is stable; from 910 °C to 1403 °C face-centred cubic γ Fe is stable, and from 1403 °C to the melting point at 1535 °C body-centred cubic δ Fe is stable. On heating α Fe its ferromagnetic properties disappear at the Curie point (768 °C) and a non-magnetic body-centred cubic structure exists from 768° to 910 °C. The loss of ferromagnetism was initially thought to be a phase change, the α Fe transforming to β Fe. Hence arose reference to α, β, γ and δ Fe. With the later experimental determination of crystal structure of the allotropes it was realised that α Fe and β Fe are both b.c.c. and differ only in their magnetic properties. It is now customary to refer to α (magnetic) and α (non-magnetic). No β phase is shown in the phase diagram, the sequence of allotropic modifications being denoted by α, γ and δ.

The change points are given individual designations,

$$\alpha \text{ (magnetic)} \quad \rightleftharpoons \alpha \text{ (non-magnetic)} \quad A2 \text{ point,}$$
$$\alpha \text{ (non-magnetic)} \rightleftharpoons \gamma \quad A3 \text{ point,}$$
$$\gamma \quad \rightleftharpoons \delta \quad A4 \text{ point.}$$

The $A1$ change point is reserved for the eutectoid reaction isotherm in Fe–C alloys. The phase diagram of Fig. 89a shows that carbon is able to dissolve to a much greater extent in γ Fe than

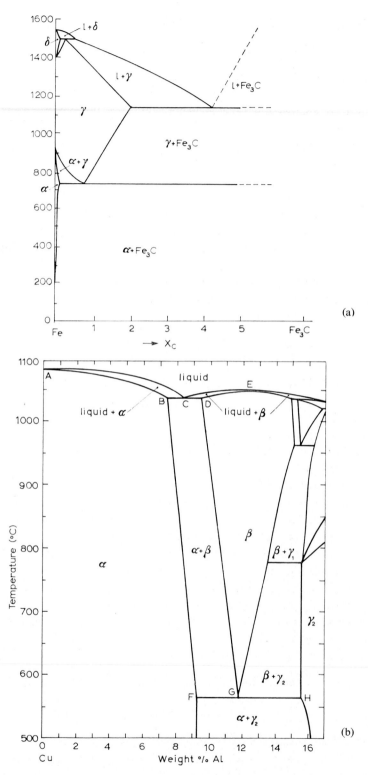

Fig. 89. Eutectoid reactions. (a) The Fe–Fe₃C system (from *Metals Handbook*, 1948, courtesy American Society for Metals); (b) a part of the Cu–Al system (courtesy Copper Development Association).

in α Fe or δ Fe. The solid solutions of carbon in α Fe, γ Fe and δ Fe are also individually named*,

the α solid solution phase is called ferrite,

the γ solid solution phase is called austenite,

the δ solid solution phase is called δ ferrite.

The phase diagram (Fig. 89a) is a partial phase diagram, with Fe and the intermediate phase Fe_3C as components. The phase Fe_3C is known as cementite. The Fe–Fe_3C partial phase diagram is a simple diagram insofar as it is built up from three invariant reaction isotherms,

(1) a peritectic reaction, $l+\delta \rightleftharpoons \gamma$

(2) a eutectic reaction, $l \rightleftharpoons \gamma + Fe_3C$

(3) a eutectoid reaction, $\gamma \rightleftharpoons \alpha + Fe_3C$.

The eutectoid reaction can be written,

austenite $\rightleftharpoons$ ferrite + cementite.

The mixture of ferrite and cementite which separates from an alloy of eutectoid composition is called pearlite. The eutectoid region is shown in more detail in Fig. 90. The eutectoid com-

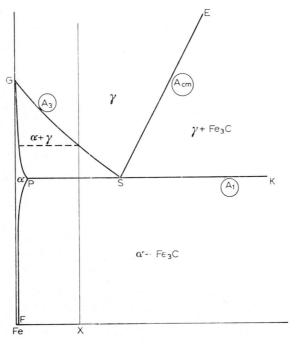

Fig. 90. The eutectoid region of the Fe–Fe_3C system.

position occurs at 0.8 wt. %C—it is usual to refer to percentage C and not to percentage Fe_3C in the Fe–Fe_3C system. The eutectoid temperature or $A1$ change point is 723 °C. There is no accepted nomenclature for such boundary lines as GS, GP and SE in a eutectoid system; as

---

* The Fe–Fe_3C system is unique in that metallurgists followed the practice of mineralogists in giving names to individual phases. The American metallurgist H. M. Howe proposed the terms ferrite, cementite and pearlite; austenite was suggested by the French metallurgist F. Osmond in memory of the English metallurgist Sir W. C. Roberts-Austen. In passing it is interesting to note that the term eutectic was used by F. Guthrie (*Phil. Mag.*, **17** (1884) 462) whilst H. M. Howe (*Metallographist*, **6** (1903) 245) proposed the word eutectoid. The German metallurgist W. Guertler coined the word peritectic (*Handbook of Metallography*, Vol. 1, Gebrüder Borntraeger, Berlin, 1912, p. 277).

might be expected in the Fe–Fe$_3$C system the $\alpha + \gamma/\gamma$ boundary *GS* is called the *A*3 line, the $\gamma/\gamma + Fe_3$C boundary *SE* is called the *Acm* line, where *cm* stands for cementite. The terms hypo-eutectoid and hyper-eutectoid are used to distinguish alloys with compositions less than, or greater than, the eutectoid alloy.

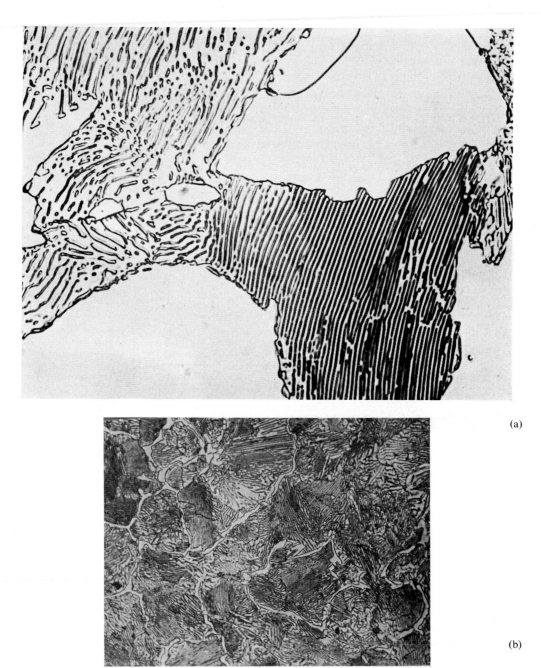

(a)

(b)

Fig. 91. Microstructure of (a) a hypo-eutectoid steel showing ferrite (white) and pearlite, $\times$1000 (courtesy The British Iron and Steel Research Association); (b) a hyper-eutectoid steel showing grain boundary cementite and pearlite grains, $\times$450 (courtesy Metallurgical Services).

A hypo-eutectoid alloy of composition X (Fig. 90) exists as γ phase (austenite) if heated to a temperature above $A3$. On cooling, this alloy meets the $A3$ line (GS) at a temperature T_1, when it begins to deposit α (ferrite). The ferrite is of much lower carbon content than the austenite from which it is deposited and therefore the composition of the austenite changes with continued cooling along the curve GS. Ferrite continues to be deposited until the temperature falls to the eutectoid temperature, $A1$. At this temperature austenite of eutectoid composition transforms to pearlite. As noted previously, this transformation tends to be sluggish. The sequence of phase separations in the eutectoid system is analogous to that in the eutectic system; pro-eutectoid ferrite is deposited over a range of temperature followed by the isothermal transformation of the remaining austenite. The pro-eutectoid ferrite is strictly comparable with the primary separation in the eutectic system. The structure of a hypo-eutectoid Fe–Fe$_3$C alloy is one of ferrite with pearlite, the latter constituent appearing in characteristic form (Fig. 91a). The lamellae of cementite act as an optical diffraction grating and give the characteristic pearly sheen to the specimen.

On heating alloy X, the structure remains one of pro-eutectoid ferrite with pearlite until $A1$ is reached. The pearlite then transforms to austenite. In any one grain the transformation of pearlite to austenite will occur at the interface between the ferrite and cementite plates of the pearlite. There is an abundant supply of nucleation points and from the original grain there will be formed many regions of austenite. From $A1$ to $A3$ the new γ regions are in equilibrium with the pro-eutectoid ferrite. As $A3$ is approached the proportion of ferrite decreases by diffusion of C atoms from the γ to the α and of Fe atoms from the α to the γ, until at $A3$ the alloy is completely austenitic. If the alloy is now cooled from $A3$ to room temperature the pro-eutectoid ferrite deposits in the more numerous new γ boundaries to produce a much finer structure than that originally produced in direct cooling from the melt.

Hyper-eutectoid alloys on cooling from the austenite phase region deposit cementite over a range of temperature until $A1$ is reached. As before, the remaining austenite then transforms to pearlite. Again the cementite which separates before $A1$ is called pro-eutectoid cementite (Fig. 91b).

One peculiarity of the Fe–Fe$_3$C phase diagram is that it is not a true equilibrium diagram. If cementite is heated at a high temperature, it will break down into austenite and graphite. Graphite formation is an extremely slow process in the binary Fe–Fe$_3$C alloys. Under certain circumstances the addition of a third element promotes graphitisation. The elements Co, Al and Si, for example, have this effect. The equilibrium between austenite and cementite and between ferrite and cementite is a metastable one; the true equilibrium is between austenite or ferrite and graphite. Therefore the Fe–Fe$_3$C phase diagram is a metastable diagram. The stable equilibrium diagram is the Fe–C diagram. Notwithstanding this statement, it should be noted that the steels (Fe–Fe$_3$Calloys) are remarkably stable materials. For all practical purposes they are regarded as being stable rather than metastable. In terms of free energy–composition curves it would seem reasonable to suppose that Fe$_3$C has a slightly higher free energy than graphite so that the common tangent between the ferrite and graphite free energy curves is only infinitesimally below that between ferrite and Fe$_3$C. The relationship between the stable Fe–C and the metastable Fe–Fe$_3$C diagrams is shown in Fig. 92. As noted previously a metastable phase has a greater solubility than a stable phase in a second phase with which they are in equilibrium (cf. p. 79). Thus cementite has a greater solubility in austenite and ferrite than has graphite. Although the eutectoid reaction has been described by reference to the Fe–Fe$_3$C system, there are many other systems which exhibit the same type of reaction.

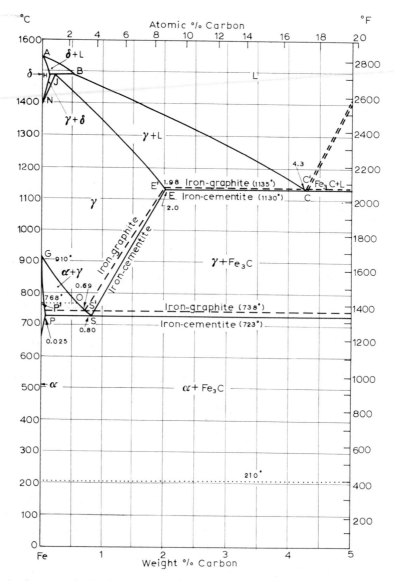

Fig. 92. Relationship between the stable Fe–C and the metastable Fe–Fe$_3$C phase diagrams; – – – – –, stable Fe–C; ————, metastable Fe–Fe$_3$C; (courtesy American Society for Metals).

The eutectoid reaction in the Cu–Al system involves the transformation of a disordered body-centred cubic intermediate phase β (based on the composition Cu$_3$Al) into the Cu-rich α solid solution and an intermediate phase γ_2, based on the composition Cu$_9$Al$_4$,

$$\beta \underset{}{\overset{565°\text{ C}}{\rightleftharpoons}} \alpha + \gamma_2.$$

Further transformation occurs at lower temperatures and this is referred to in section 6.4.

6.3. MONOTECTOID REACTIONS

As in the case of eutectoid reactions a monotectoid reaction can occur with or without an allotropic transformation in one of the components. In the former case the components usually form a complete series of solid solutions at high temperatures—for example the β phase in the Ta–Zr

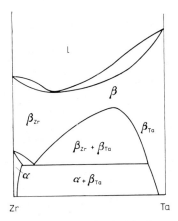

Fig. 93. The monotectoid reaction in the Ta–Zr system (schematic).

system* shown in Fig. 93. At lower temperatures β_{Zr} breaks down into the α phase and β_{Ta},

$$\beta_{Zr} \rightleftharpoons \alpha + \beta_{Ta}.$$

Both β_{Zr} and β_{Ta} have the same crystal structure (b.c.c.) but different lattice spacing.

The monotectoid reaction in the Al–Zn system is an example of a reaction without an allotropic transformation of a component. The α solid solution (Fig. 94a) has a very wide range of stability at high temperatures but undergoes the monotectoid reaction at lower temperatures,

$$\alpha' \rightleftharpoons \alpha + \beta.$$

Both α and α' are face-centred cubic phases, differing only in lattice spacing. Hume-Rothery

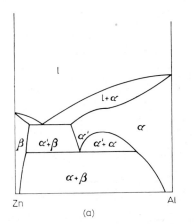 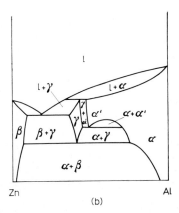

Fig. 94. The monotectoid reaction in the Al–Zn system (schematic). (a) Previously accepted phase diagram; (b) recently proposed modification.

---

* D. E. WILLIAMS, R. J. JACKSON AND W. L. LARSEN, *Trans. Met. Soc. AIME*, **224** (1962) 751.

and Raynor* have considered the formation of a monotectoid in the Al–Zn system with reference to hypothetical free energy curves for the α and β phases and to electronic factors. However, recent work** indicates a peritectic reaction at 443 °C with an extremely narrow $\alpha' + \gamma$ phase region falling to the monotectoid horizontal at 340 °C. The monotectoid reaction in this case requires the decomposition of the α' phase to give a mixture of α and γ (Fig. 94b).

6.4. PERITECTOID REACTIONS***

Recent work† on Cu–Al alloys with compositions within the $\alpha + \gamma_2$ phase region (Fig. 89b) has indicated that a peritectoid reaction may occur at a temperature 200 °C below the eutectoid isothermal between the α and γ_2 phases, with the production of an intermediate phase α_2,

$$\alpha + \gamma_2 \overset{363° \text{ C}}{\rightleftharpoons} \alpha_2.$$

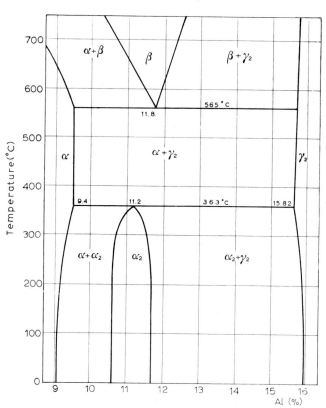

Fig. 95. Revision of the Cu–Al phase diagram by R. P. Jewett and D. J. Mack to indicate a peritectoid reaction at 363 °C. (After R. P. JEWETT AND D. J. MACK, *J. Inst. Met.*, **92** (1963–64) 59; courtesy Institute of Metals).

　　* W. HUME-ROTHERY AND G. V. RAYNOR, *The Structure of Metals and Alloys*, Monograph No. 1, 4th edn., Institute of Metals, London, 1962, p. 137.

　** G. R. GOLDAK AND J. GORDON PARR, *J. Inst. Metals*, **92** (1963–64) 230; A. A. PRESNYAKOV, YU. A. GORBAN AND V. V. CHERVOYAKOVA, *Zh. Fiz. Khim.*, **35** (1961) 1289; *Tr. Inst. Yadern. Fiz., Akad. Nauk Kaz. SSR,* (4) (1961) 85.

*** Peritectoid reactions are occasionally referred to as metatectic reactions. The term metatectic reaction has been used (p. 43) to define an equilibrium of the type $\alpha \rightleftharpoons \beta + l$. It is contrary to accepted nomenclature to use the term meta*tectic* for a solid state reaction).

　† D. R. F. WEST AND D. LLOYD THOMAS, *J. Inst. Metals*, **83** (1954–55) 505; R. P. JEWETT AND D. J. MACK, *ibid.*, **92** (1963–64) 59; J. S. LLEWELYN LEACH, *ibid.*, **92** (1963–64) 93.

This peritectoid reaction is claimed to be very sluggish, and heat treatment times up to 14 weeks were necessary to produce the reaction. The proposed revision to the phase diagram is shown in Fig. 95.

The system Ag–Al* contains two peritectoid reactions in addition to two peritectic and a eutectic reaction. To illustrate how a complex diagram of this type can be built up from the basic building blocks—the invariant reaction isothermals—reference should be made to Fig. 96. The invariant reactions are drawn as full lines and they have subsequently been linked

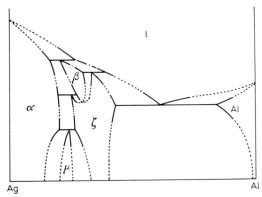

Fig. 96. The Ag–Al phase diagram (schematic). (After W. HUME-ROTHERY AND G. V. RAYNOR; courtesy Institute of Metals.)

together by dotted lines. Given the reaction $\beta + \alpha \rightleftharpoons \zeta$ is a peritectoid, there must be a minimum in the $\beta + \zeta$ phase region. The diagram can be completed by designating the two-phase fields.

Some puzzling features in complex equilibrium diagrams can be resolved if they are con-

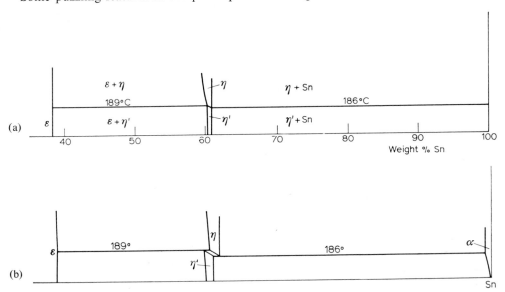

Fig. 97. (a) A part of the Cu–Sn phase diagram (after G. V. RAYNOR; courtesy Institute of Metals); (b) equilibrium relationships if the data in (a) are considered limiting cases of the peritectoid and eutectoid reactions.

* H. W. L. PHILLIPS, *Institute of Metals Annotated Equlibrium Diagrams*, No. 21, 1956.

sidered to be limiting cases of the eutectoid and peritectoid reactions. In Fig. 97a some reaction occurs at 189 °C for alloys whose compositions cover the range 38.25% to 60.3% Sn; alloys within the composition range from 60.9% to >99.99% Sn undergo a reaction at a slightly lower temperature of 186 °C. A possible explanation of the data is given in Fig. 97b. The reaction at 189 °C can be considered a limiting case of the peritectoid reaction, and the reaction at 186 °C to be a limiting case of the eutectoid reaction. Figure 97b obeys the Phase Rule, in contrast to the construction shown in Fig. 97a.

Since recent work* has confirmed that the $\eta \to \eta'$ transition is an order–disorder transformation, the interpretation of the equilibria presented in Fig. 97b conforms with the treatment of order–disorder transformations as classical phase changes obeying the Phase Rule.

---

* G. J. DAVIES, *J. Inst. Metals*, **93** (1964–65) 197.

CHAPTER 7

Binary Phase Diagrams.
Allotropy of the Components

Several commercially important metals exist in more than one crystalline form. Iron has three allotropes—α, γ and δ. Titanium has two allotropes—close-packed hexagonal α Ti stable at low temperature and body-centred cubic β Ti stable at high temperature. Zirconium behaves similarly. Plutonium has the highest number of modifications known—six.

Taking the usual case where either two or three allotropic modifications exist for a particular component, the binary phase diagrams formed can be classified into two broad groups:

   (1) diagrams in which one phase is in equilibrium with the liquid phase;

   (2) diagrams in which two phases are in equilibrium with the liquid phase.

7.1. SYSTEMS IN WHICH ONE PHASE IS IN EQUILIBRIUM WITH THE LIQUID PHASE

Such systems can be further divided according to whether the high-temperature phase forms a continuous series of solid solutions with the other component or not.

7.1.1. The high-temperature phase forms a series of solid solutions with the other component

In addition to the monotectoid system (Fig. 93) there are two other basic types of system as shown in Fig. 98a and b. Figure 98c and d illustrates the equilibria found when both components have two allotropic modifications.

The transformations which occur on cooling alloys from the β phase region to the α phase region in systems of the type shown in Fig. 98a (for example the Ti–Ta and Ti–Mo systems) are analogous to those discussed for the solidification of alloys in systems with continuous series of solid solutions. The loop-shaped phase region illustrated in Fig. 98b has not been

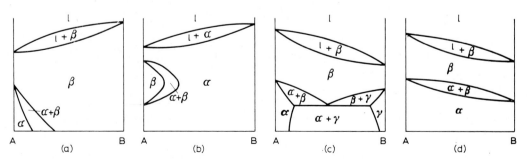

Fig. 98. Types of phase diagrams formed when the high temperature allotrope forms a continuous series of solid solutions with the second component.

considered so far. On cooling alloys such as X (Fig. 99) from the α phase region the lever rule indicates that the amount of β phase precipitated increases from temperature T_1 to T_2 (the temperature for maximum solubility of component B in the β phase) and then decreases from T_2 to T_3. At T_3 the alloy is once again completely α solid solution. Alloy Y when cooled from the α phase region is transformed completely to β at high temperature, but transforms back to α at lower temperatures. This type of system is common where one component has three allotropes—examples are found in the Fe–Cr and Fe–V systems*.

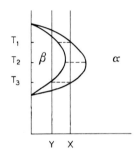

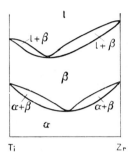

Fig. 99. Cooling of alloys through the β loop.   Fig. 100. The Ti–Zr phase diagram (schematic).

The system illustrated in Fig. 98c has a eutectoidal decomposition of the high-temperature allotrope β into α and γ, the low-temperature allotropes of components A and B respectively. Complete series of solid solutions are formed between each of the allotropes in the system Ti–Zr. This system is similar to that shown in Fig. 98d except that there is a minimum in the liquidus and in the $\alpha + \beta$ phase region (Fig. 100).

7.1.2. *Both phases form limited solid solutions with the other component*

In such systems the high-temperature β phase, as well as the low-temperature α phase, form limited solid solutions with component B. Examples are given in Fig. 101. It should be noted that Fig. 101a is the portion of Fig. 101b above a temperature T. If the eutectoid in Fig. 101b were replaced by a peritectoid the diagram would be similar to the Zr–Gd system**.

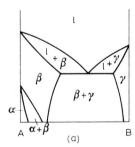   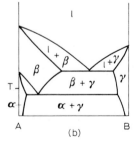   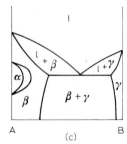

Fig. 101. Examples of phase diagrams formed when the second component has limited solubility in both allotropes.

---

   * Both systems have an intermediate phase called the σ phase, which forms in the solid state, with a composition based on FeCr and FeV respectively (*cf.* Fig. 33).
  ** M. I. COPELAND, C. E. ARMANTROUT AND H. KATO, *U.S. Bur. Mines, Rept. Invest. No. 5850* (1961).

7.2. SYSTEMS IN WHICH TWO PHASES ARE IN EQUILIBRIUM WITH THE LIQUID PHASE

A selection of systems of interest is given in Fig. 102. In the first two systems $l+\alpha$ and $l+\beta$ phase regions appear. In the remaining systems α is not in equilibrium with the liquid, but $l+\beta$ and $l+\gamma$ phase regions are formed.

A simple peritectic reaction occurs in Fig. 102a. According to Köster and Horn*, the systems Co–Os, Co–Re and Co–Ru are of this type. The reaction in Fig. 102b is an example of a metatectic reaction,

$$\beta \rightleftharpoons \alpha + l.$$

Since this is the first time this reaction has been encountered, the system will be considered in some detail. A similar reaction is found in Fig. 102f, where $\gamma \rightleftharpoons \beta + l$. The relevant portion of

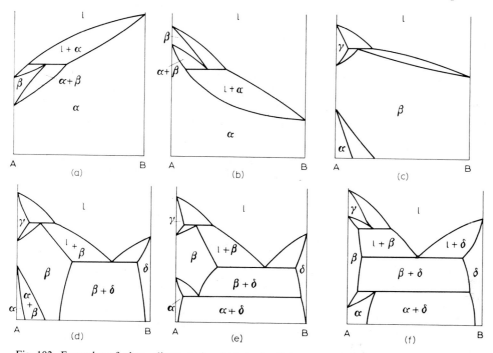

Fig. 102. Examples of phase diagrams in which both allotropes are in equilibrium with the melt.

the phase diagram is repeated in Fig. 103. Alloys with compositions from A to a solidify initially as β solid solution, but undergo transformation at lower temperatures to α. Alloys with the metatectic composition M precipitate β on cooling from the liquid state. The amount of β increases with further cooling until the alloy is 100% β, i.e. completely solid, by the time it reaches the metatectic temperature. However, alloy M will only be completely solid for a short time because, at T_M, the β undergoes the metatectic reaction,

$$\beta_M \rightleftharpoons \alpha_a + l_b.$$

As a result, alloy M partially re-melts. Further cooling leads to the separation of increasing amounts of α from the liquid until the alloy eventually freezes as 100% α. Alloys with composi-

* W. Köster and E. Horn, Z. Metallk., 43 (1952) 444.

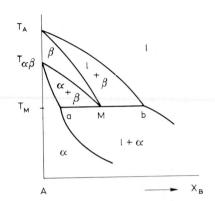

Fig. 103. The metatectic reaction.

tions between a and M solidify as β, from which some α is precipitated as the temperature falls to T_M. Again at T_M these alloys partially re-melt. The alloys between M and b also re-melt partially, *i.e.* at T_M the amount of liquid already present is increased as a result of the metatectic reaction. It is the alloys between a and M which undergo the most startling change in that they solidify completely, remain solid for a considerable temperature interval, and then partially re-melt at T_M. The schematic free energy–composition curves for this system are given in

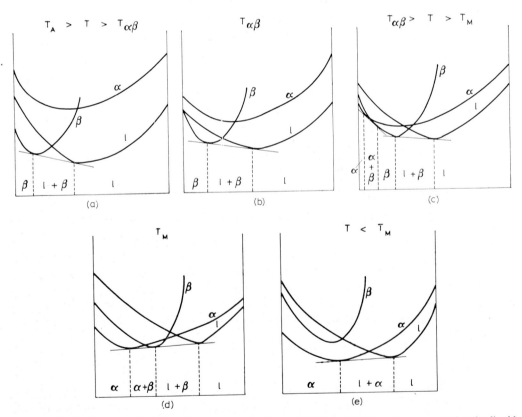

Fig. 104. Derivation of the metatectic phase diagram (Fig. 103) from the free energy curves for the liquid, α and β phases.

Fig. 104. As the temperature falls the β curve rises in relation to the liquid and α curves, so that the equilibrium between β and liquid is eventually replaced by the $l+\alpha$ equilibrium.

A metatectic reaction also appears in the system shown in Fig. 102f. An example of this type of system is the partial phase diagram Fe–Fe$_2$Zr, where $\delta \rightleftharpoons \gamma + l$ at 1335 °C. Of the remaining systems, that in Fig. 102e is similar to the Fe–Fe$_3$C system and Fig. 102c to the Fe–Ni system.

CHAPTER 8

Ternary Phase Diagrams.
Two-Phase Equilibrium

8.1. INTRODUCTION

A system with three components is called a ternary system. A ternary alloy system is generally composed of three metals A, B and C. Partial ternary systems, in which intermediate phases are considered as components of the system, are also frequently met.

Taking the condensed Phase Rule, $p+f = c+1$, the equilibria possible in a ternary system are summarised below.

Number of components	Number of phases	Variance	Equilibrium
$c = 3$	$p = 1$	$f = 3$	trivariant
$c = 3$	$p = 2$	$f = 2$	bivariant
$c = 3$	$p = 3$	$f = 1$	monovariant
$c = 3$	$p = 4$	$f = 0$	invariant

Trivariant equilibrium: $p = c-2$

If only one phase is present in a ternary system there are three degrees of freedom. The case of trivariant equilibrium therefore allows one to select the temperature and the concentration of two components in the phase (this defines the composition of the ternary phase since $X_A + X_B + X_C = 1$) in an arbitrary manner. No other variables are required to specify the state of any alloy in a one-phase region of a ternary system. It follows that the temperature and the composition of any alloy can be varied separately or together with no change in the state of the system (within the limits of the one-phase region).

One-phase equilibrium corresponds to the existence of distinct liquid (l_1, l_2, etc.) or solid (α, β, γ, etc.) phase regions in the ternary. With three variables required to specify the state of a ternary one-phase region, it is obvious that a three-dimensional diagram is necessary to represent ternary equilibria.

Bivariant equilibrium: $p = c-1$

According to the Phase Rule a ternary system with two-co-existing phases will have two degrees of freedom. This implies that either

(a) the temperature and the concentration of one component in one of the phases, or

(b) the concentration of two of the components in one of the phases, or

(c) the concentration of one component in one phase and the concentration of another component in the second phase, may be arbitrarily selected. Once the selection has been made this establishes

(a) the composition of the second phase,

(b) the temperature and composition of the second phase,

(c) the temperature and composition of the two phases, respectively.

The bivariant equilibria possible in ternary systems are $l_1 \rightleftharpoons l_2$, $l \rightleftharpoons \alpha$, and $\alpha \rightleftharpoons \beta$.

Monovariant equilibrium: p = c

With three phases in equilibrium a ternary system has one degree of freedom. In this case only one of the external variables can be freely chosen. Once this variable has been selected the other variables are automatically fixed. If the temperature is chosen arbitrarily, the compositions of the three phases are fixed. On the other hand, if the concentration of one component in one of the phases is selected, the temperature and compositions of all three phases are automatically fixed.

The monovariant reactions which can take place in a ternary system are:

$$\alpha \rightleftharpoons \beta + \gamma, \qquad \alpha \rightleftharpoons \beta + l, \qquad \alpha \rightleftharpoons l_1 + l_2$$
$$l_1 \rightleftharpoons l_2 + l_3, \qquad l_1 \rightleftharpoons \alpha + l_2, \qquad l \rightleftharpoons \alpha + \beta$$
$$\alpha + \beta \rightleftharpoons \gamma, \qquad \alpha + \beta \rightleftharpoons l, \qquad l_1 + l_2 \rightleftharpoons l_3$$
$$l_1 + l_2 \rightleftharpoons \alpha, \qquad l_1 + \alpha \rightleftharpoons l_2, \qquad l + \alpha \rightleftharpoons \beta.$$

Invariant equilibrium: p = c+1

A ternary four-phase equilibrium has no degrees of freedom. Such a system can only exist at one fixed temperature and the composition of all four phases is fixed. The invariant reactions which are theoretically possible in a ternary system are:

$$\alpha \rightleftharpoons \beta + \gamma + \delta, \qquad \alpha + \beta \rightleftharpoons \gamma + \delta, \qquad \alpha + \beta + \gamma \rightleftharpoons \delta$$
$$l_1 \rightleftharpoons l_2 + l_3 + l_4, \qquad l_1 + l_2 \rightleftharpoons l_3 + l_4, \qquad l_1 + l_2 + l_3 \rightleftharpoons l_4$$
$$l \rightleftharpoons \alpha + \beta + \gamma, \qquad l + \alpha \rightleftharpoons \beta + \gamma, \qquad l + \alpha + \beta \rightleftharpoons \gamma$$
$$l_1 \rightleftharpoons l_2 + \alpha + \beta, \qquad l_1 + l_2 \rightleftharpoons \alpha + \beta, \qquad l_1 + l_2 + \alpha \rightleftharpoons \beta$$
$$l_1 \rightleftharpoons l_2 + l_3 + \alpha, \qquad l_1 + l_2 \rightleftharpoons l_3 + \alpha, \qquad l_1 + l_2 + l_3 \rightleftharpoons \alpha$$
$$\alpha \rightleftharpoons l_1 + l_2 + l_3, \qquad \alpha + l_1 \rightleftharpoons l_2 + l_3, \qquad \alpha + l_1 + l_2 \rightleftharpoons l_3$$
$$\alpha \rightleftharpoons \beta + l_1 + l_2, \qquad \alpha + \beta \rightleftharpoons l_1 + l_2, \qquad \alpha + \beta + l_1 \rightleftharpoons l_2$$
$$\alpha \rightleftharpoons \beta + \gamma + l, \qquad \alpha + \beta \rightleftharpoons \gamma + l, \qquad \alpha + \beta + \gamma \rightleftharpoons l$$
$$l_1 + \alpha \rightleftharpoons l_2 + \beta.$$

The symbols l_1, l_2, l_3, l_4 denote four distinct liquid phases and α, β, γ, δ four solid phases. According to Findlay* it is an empirical observation that the number of liquid phases never exceeds the number of components in the system. On this basis the three reactions involving four liquid phases would be ruled out.

In a ternary system a four-phase equilibrium corresponds to the combination of four invariant points. It lies on a horizontal (isothermal) plane.

8.2. REPRESENTATION OF TERNARY SYSTEMS

The maximum number of independent variables in a ternary system at constant pressure is three—the temperature and two concentration values. To represent ternary equilibria a three-dimensional phase diagram is necessary. It is usual to draw such diagrams with the composition represented in a horizontal plane and the temperature on an axis perpendicular to this plane.

---

* A. FINDLAY, in A. N. CAMPBELL AND N. O. SMITH (eds.), *The Phase Rule and Its Applications*, Dover Publications Inc., New York, 9th edn., 1951, p. 8.

The composition of a ternary phase α is established by selecting two variables, X_A^α and X_B^α, which specify the concentrations of components A and B in the α phase. The concentration of the third component, X_C^α, is obtained from the relation

$$X_A^\alpha + X_B^\alpha + X_C^\alpha = 1.$$

The concentrations are then expressed as weight, atomic or mole (g.-atom) fractions. If the relation $X_A^\alpha + X_B^\alpha + X_C^\alpha = 100$ is used, then the concentrations are expressed in weight, atomic or mole percentages.

The most common method for plotting composition in a ternary system uses an equilateral triangle*, sometimes referred to as the Gibbs triangle (Fig. 105). The corners represent the pure components A, B and C. The edges AB, AC and BC represent the corresponding binary compositions. Ternary compositions are plotted as points within the triangle ABC.

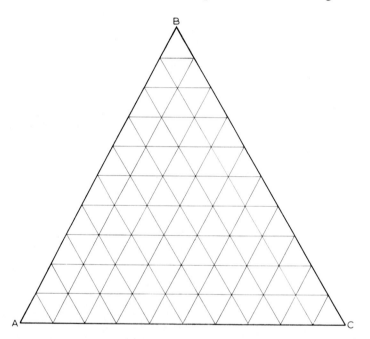

Fig. 105. The Gibbs triangle.

In an equilateral triangle the sum of the perpendiculars from any point in the triangle to the sides of the triangle are constant and equal to the height of the triangle. The perpendicular distances may be used to give the composition of any alloy in the ternary system. In triangle ABC (Fig. 106) perpendiculars AO_1, BO_2 and CO_3 from the corners, have each been divided into ten equal parts. If a series of lines are now drawn through the divisions such that they are

---

* *The Scientific Papers of J. Willard Gibbs*, Vol. 1, *Thermodynamics*, Dover Publications Inc., New York, 1961, p. 118. A re-issue of the original 1906 edition. The work was initially published in *Trans. Conn. Acad.*, **3** (1876) 176. As with much of Gibbs' work, the publication was unnoticed and the method was subsequently re-discovered by Sir G. G. STOKES (*Proc. Roy. Soc.* (*London*), **49** (1891) 174), who adapted Maxwell's colour triangle for the purpose. It is an interesting footnote to history to note that Stokes' paper was called a "Note on a Graphical Representation of the Results of Dr. Alder Wright's Experiments on Ternary Alloys" and was an interpretation of the first published work on ternary alloy phase equilibria (C. R. A. WRIGHT AND C. THOMPSON, Alloys of Pb, Sn and Zn, *Proc. Roy. Soc.* (*London*), **45** (1889) 461).

at right angles to the perpendiculars, these lines will represent compositions at a constant distance from an edge of the triangle. The line ac is at right angles to the perpendicular BO_2 and all points on ac are at a distance 0.1 from the edge AC. The distance of the line ac from the edge AC represents the fraction of component B (*i.e.* that component which is opposite to the edge in question) in alloys along the line ac. In this case there is a fraction 0.1 of B in alloys along ac. If the height of the triangle were assumed to be 100, then alloys along ac would contain 10% B.

An alloy P lies on lines distant 0.3 from BC, 0.2 from AC and 0.5 from AB. The fractions of A, B and C in alloy P are equal to the perpendicular distances PP_1, PP_2 and PP_3 respectively. Alloy P therefore contains 0.3 A, 0.2 B and 0.5 C, or 30% A, 20% B and 50% C. Alloys with compositions on one edge of the triangle, as alloy Q, contain no B since the perpendicular distance from Q to the edge AC is zero. Alloys between A and C are therefore binary; alloy

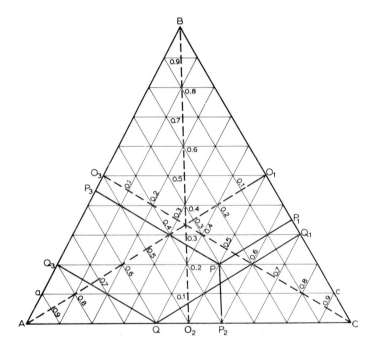

Fig. 106. Plotting of alloy compositions in ternary systems.

Q contains 40% C and 60% A. Compositions at the corners of the triangle obviously represent pure components since only one perpendicular can be dropped from a corner. For corner A this is perpendicular AO_1.

We have noted that all points on a line drawn parallel to one side of the triangle ABC contain a constant amount of the component opposite to the edge. Similarly, all points on lines which are drawn from a corner to the opposite edge of the triangle contain a constant ratio of the components along the edge. Alloys on the line BO_2 contain a constant ratio of A : C. It is immaterial whether the line meets the edge AC at right angles or not, as simple geometrical consideration will show.

To represent a ternary system completely the effect of temperature has to be considered. By plotting the temperature on an axis perpendicular to the composition triangle a right-triangular

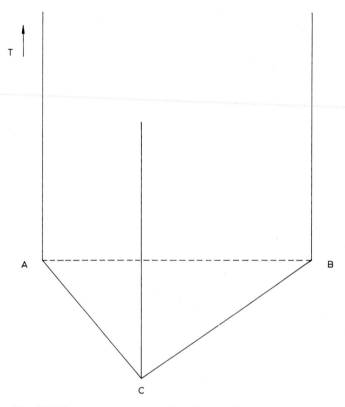

Fig. 107. Temperature–concentration diagram for a ternary system.

prism (Fig. 107) results. The prism edges represent the effect of temperature on the components of the system, the sides of the prism correspond to the three binary T–X phase diagrams and the interior represents the ternary space diagram.

8.3. TIE LINES AND TIE TRIANGLES

In a binary system the lever rule indicates the relative amounts (in terms of atom fractions, weight fractions, atomic or weight percentages) of the two phases in a phase mixture in terms of the composition of the alloy and the phases into which it separates. The same rule can be applied to any ternary alloy X which separates into two phases (Fig. 108) α and β,

$$\frac{m_\beta}{m_\alpha} = \frac{\alpha X}{X\beta}.$$

That this is so can be seen by considering the binary system AC (or BC) compared with the ternary:

$$\frac{m_\beta}{m_\alpha} = \frac{X_A - X_A^\alpha}{X_A^\beta - X_A} = \frac{ab}{b\beta} = \frac{\alpha X}{X\beta}.$$

The line $\alpha\beta$ is called a tie line.

Tie triangles occur in ternary systems when three phases co-exist (Fig. 109). Assume that the

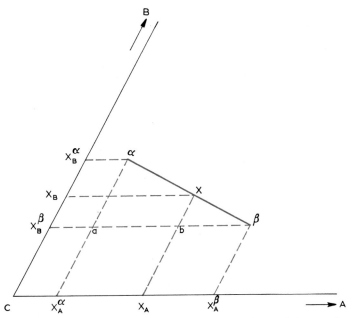

Fig. 108. Application of the lever rule to a ternary alloy X consisting of the two phases α and β.

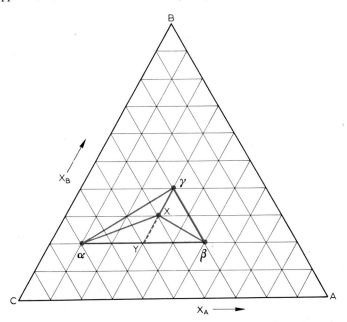

Fig. 109. Application of the lever rule to a ternary alloy X consisting of the three phases α, β and γ.

proportions of $\alpha : \beta : \gamma$ are in the ratio $0.25 : 0.25 : 0.5$. The co-ordinates of the phases are:

α	10% A,	20% B,	70% C,
β	50% A,	20% B,	30% C,
γ	30% A,	40% B,	30% C.

Applying the lever rule to the α and β phases we see that these phases are formed from an alloy Y of composition halfway between α and β (30% A, 20% B, 50% C) since $m_\alpha = m_\beta$. Now we can apply the lever rule again to the γ phase and alloy Y. By this means we find that Y and γ are formed from alloy X, which is midway between Y and γ since $m_\alpha + m_\beta = m_Y = m_\gamma$. Point X can be considered the centre of gravity of triangle $\alpha\beta\gamma$ at whose vertices masses m_α, m_β and m_γ are placed. These masses are related, $m_\alpha : m_\beta : m_\gamma = 0.25 : 0.25 : 0.5$. The co-ordinates of point X are 30% A, 30% B, 40% C. The triangle $\alpha\beta\gamma$ is referred to as a tie triangle. It indicates the compositions of co-existing phases in a three-phase equilibrium.

The composition of X, the fulcrum of the tie triangle, could have been obtained directly from the compositions and relative amounts of α, β and γ:

$$\left. \begin{array}{l} 0.25\,(10\%) + 0.25\,(50\%) + 0.5\,(30\%) = 30\% \text{ A} \\ 0.25\,(20\%) + 0.25\,(20\%) + 0.5\,(40\%) = 30\% \text{ B} \\ 0.25\,(70\%) + 0.25\,(30\%) + 0.5\,(30\%) = 40\% \text{ C} \end{array} \right\} X$$

It is appropriate to consider at this stage an extension of the tie triangle discussion to the three cases where four phases co-exist. Taking the first case the point X (Fig. 109) would now be considered the fourth phase. If the phase within the triangle $\alpha\beta\gamma$ is a liquid phase, we can regard this liquid phase as being in equilibrium with the α, β and γ phases,

$$l \rightleftharpoons \alpha + \beta + \gamma$$

where $m_l = m_\alpha + m_\beta + m_\gamma$ (Fig. 110a). The proportions of the α, β and γ phases which form from the liquid phase l are in the proportions of the areas of triangles $l\beta\gamma$, $l\alpha\gamma$ and $l\alpha\beta$ respectively. The above reaction is called, by analogy with its binary counterpart, a ternary eutectic reaction. It will be seen later that such a reaction involves four tie triangles which meet on the four-phase plane $\alpha\beta\gamma l$; tie triangles $l\alpha\gamma$, $l\alpha\beta$ and $l\beta\gamma$ descending from above this plane, and tie triangle $\alpha\beta\gamma$ descending from the plane to lower temperatures.

The other two cases of four-phase equilibrium are somewhat analogous to the peritectic reaction in binary systems. In both cases the point representing the composition of the liquid phase lies outside the triangle $\alpha\beta\gamma$. In Fig. 110b the liquid composition is opposite the α corner; in Fig. 110c the liquid composition is opposite the $\alpha\beta$ edge of the $\alpha\beta\gamma$ triangle. It will be obvious that the latter phase distribution allows the reaction

$$l + \alpha + \beta \rightleftharpoons \gamma.$$

One tie triangle $l\alpha\beta$ descends from high temperature to the four-phase plane and the three tie triangles $l\alpha\gamma$, $l\beta\gamma$ and $\alpha\beta\gamma$ descend from plane $l\alpha\beta\gamma$ to lower temperatures.

In Fig. 110b the composition X can be obtained in two ways—either by mixing α and l or

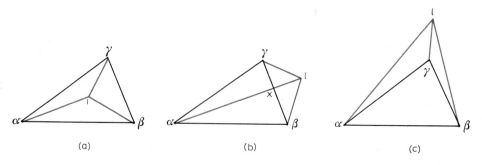

Fig. 110. Four-phase equilibrium, (a) $l \rightleftharpoons \alpha + \beta + \gamma$, (b) $l + \alpha \rightleftharpoons \beta + \gamma$, (c) $l + \alpha + \beta \rightleftharpoons \gamma$.

β and γ. Taking the relative masses of the four phases as m_α, m_β, m_γ and m_l the lever rule indicates that

$$\frac{m_\alpha}{m_l} = \frac{Xl}{\alpha X} \quad \text{and} \quad \frac{m_\beta}{m_\gamma} = \frac{\gamma X}{X\beta}.$$

Point X represents the fulcrum of the quadrangle $\alpha\beta l\gamma$ at whose vertices are placed masses m_α, m_β, m_γ and m_l. A reaction of the type

$$l + \alpha \rightleftharpoons \beta + \gamma$$

occurs with the phase distribution given in Fig. 110b. The tie triangles $l\alpha\gamma$ and $l\alpha\beta$ descend from high temperature to the four-phase plane $l\alpha\beta\gamma$ and the tie triangles $\alpha\beta\gamma$ and $l\beta\gamma$ descend from plane $l\alpha\beta\gamma$ to lower temperatures.

8.4. TWO-PHASE EQUILIBRIUM.
COMPLETE SOLUBILITY OF THE COMPONENTS IN BOTH THE LIQUID AND SOLID STATE

By analogy with eqn. (97) for the excess free energy of a binary solution we may write for a ternary solution

$$\Delta G_\mathrm{m} = aX_AX_B + bX_AX_C + cX_BX_C + NkT[X_A \ln X_A + X_B \ln X_B + X_C \ln X_C]$$

where

$$a = zN\left[H_{AB} - \frac{H_{AA} + H_{BB}}{2}\right]$$

$$b = zN\left[H_{AC} - \frac{H_{AA} + H_{CC}}{2}\right]$$

$$c = zN\left[H_{BC} - \frac{H_{BB} + H_{CC}}{2}\right].$$

As in the binary systems, the interplay of the enthalpy terms and the entropy terms will determine the type of equilibrium. As an example, if $a = b = c = 0$, the three components form ideal solutions. This is the simplest case of a ternary two-phase equilibrium between a liquid and a solid solution. There are three possible types of two-phase equilibrium:

$$l \rightleftharpoons \alpha, \qquad \alpha_1 \rightleftharpoons \alpha_2, \qquad \text{and} \qquad l_1 \rightleftharpoons l_2.$$

8.4.1. *Two-phase equilibrium between the liquid and a solid solution:* $l \rightleftharpoons \alpha$

According to the Phase Rule a one-phase region, such as the liquid or α phase region, is trivariant and is therefore represented in the ternary equilibrium diagram by a volume or state space. There is only one possible equilibrium between the liquid and α phase—the bivariant $l \rightleftharpoons \alpha$ equilibrium representing the solidification of α from the melt on cooling. A bivariant equilibrium has two degrees of freedom and is represented in the ternary space diagram* by two-dimensional geometrical elements (surfaces). Each phase will be represented by a surface. In Fig. 111 the upper surface is called the liquidus surface, and the lower surface is

---

* The term space diagram refers to the temperature–concentration diagram for the ternary system.

called the solidus surface. Above the liquidus surface all alloys within the system are molten; below the solidus surface all alloys are solid. Between the two surfaces all alloys are two-phase mixtures of liquid $+\alpha$.

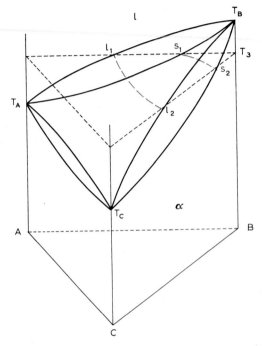

Fig. 111. Space model of a ternary system with continuous series of liquid and solid solutions.

It is appropriate at this stage to consider the methods commonly used to describe and evaluate reactions in ternary systems. These can be summarised as:

    (1) three-dimensional models;

    (2) projections on the concentration triangle ABC of

        (a) the liquidus surface,

        (b) the solidus surface,

        (c) the solid solubility surfaces;

    (3) isothermal sections;

    (4) vertical sections;

    (5) polythermal projections on the concentration triangle of the complete three-dimensional space diagram.

In the present system the three-dimensional model (Fig. 111) is easy to draw and interpret. However, as the complexity of the system increases so do the difficulties of adequately representing such systems by clear three-dimensional models. Frequently, projections of the various surfaces—liquidus, solidus, and solid solubility—on the base of the three-dimensional model are found in the literature. This base is generally assumed to be a room temperature isothermal section and is commonly called the concentration triangle. Figure 112 shows the projection of the liquidus and solidus surfaces of the system represented in Fig. 111 in the form of a number of isothermal sections through the diagram. Such sections are best taken at fixed temperature intervals and they can be used in the same way as the height contours on ordnance

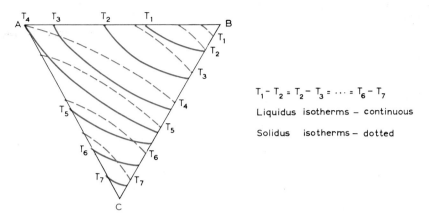

$$T_1 - T_2 = T_2 - T_3 = \cdots = T_6 - T_7$$

Liquidus isotherms — continuous

Solidus isotherms — dotted

Fig. 112. Projection of the liquidus and solidus of Fig. 111 on the concentration triangle ABC.

survey maps. A close spacing of the isotherms indicates a steep slope, whereas a wide spacing indicates a gentle slope of the relevant surface. Projections of the liquidus surface are often useful in conveying a clear impression of the shape of the surface and indicating, by folds and valleys, the presence of ternary invariant reactions.

Isothermal sections

Isothermal (horizontal) sections through the space diagram are the most widely used method for presenting experimental data. An isothermal section at temperature T_3 is shown in the form of dotted lines in Fig. 111 so as not to confuse the three-dimensional representation of the system. Figure 113 shows this section in greater detail. Where the section intersects the liquidus surface a curved line l_1l_2 is generated; a similar curved line s_1s_2 is generated at the intersection with the solidus surface. To the left of the line l_1l_2 all alloys are molten at T_3 (in the area $l_1l_2CAl_1$); to the right of the line s_1s_2 all alloys are completely solid at T_3 (in the area $s_1s_2Bs_1$). Between the liquidus line l_1l_2 and the solidus line s_1s_2 all alloys are two-phase, the liquid on line l_1l_2 co-existing with the α solid solution on line s_1s_2. The compositions of the co-existing liquid and solid phases are given by means of tie lines such as l_3s_3. It will be recalled that the horizontal line l_1s_1 in the binary system AB joins the compositions of phases in equilibrium with each other (so-called conjugate phases) at T_3. The tie lines in the ternary system (Fig. 113) joining points on the line l_1l_2 to points on line s_1s_2 are similar to the binary tie line l_1s_1.

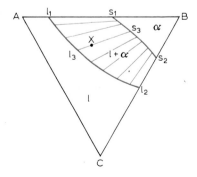

Fig. 113. Isothermal section through Fig. 111 at a temperature T_3 where $T_B > T_3 > T_A$.

Thus the tie line $l_3 s_3$ joins the conjugate phases l_3 and s_3 and indicates that, at T_3, any alloy whose composition is represented by a point on the tie line $l_3 s_3$ will consist of a liquid phase l_3 in equilibrium with a solid phase s_3. The lever rule can be applied to determine the relative amounts of l_3 and s_3. For alloy X on tie line $l_3 s_3$,

$$\frac{m_\alpha}{m_l} = \frac{l_3 X}{X s_3}$$

where m_l and m_α are the amounts of the liquid and α phase. An alloy whose composition is represented by point l_3 would contain no solid phase whatsoever since l_3 and X would be identical points. This condition corresponds to the state of affairs immediately prior to crystallisation of some α phase from the liquid and will be referred to later when considering transformations in individual alloys.

Certain general principles can be stated with regard to tie lines in ternary systems.

(1) If a two-phase region stretches from one binary to another (Fig. 113), there is a gradual transition from one binary tie line to the other binary tie line through the ternary system at a given temperature. The tie lines in Fig. 113 gradually change their slope from the side AB to the side BC of the concentration triangle.

(2) In general, only the binary tie lines when produced intersect a corner of the concentration triangle. The ternary tie lines, if produced, would not intersect at a corner of the triangle. An exception to this rule occurs when a component exhibits extremely small solid solubility and the tie lines join points on the liquidus to a corner of the concentration triangle (Fig. 114).

(3) At a given temperature tie lines cannot intersect. The consequences of intersection can be shown to be contrary to the Phase Rule in the following manner. Assume tie lines $l_5 s_5$ and $l_6 s_6$ intersect (Fig. 115). Assume also that the tie line $l_4 s_4$ meets the tie lines $l_6 s_6$ at point $s_{4(6)}$ on the solidus line $s_1 s_2$. Then, considering tie lines $l_4 s_4$ and $l_6 s_6$, three phases are in equilibrium at a constant temperature. According to the Phase Rule this equilibrium of liquid phase l_4, l_6 and solid $s_{4(6)}$ is monovariant, whereas the equilibrium under discussion is bivariant by original definition. Hence the tie lines cannot intersect at a constant temperature.

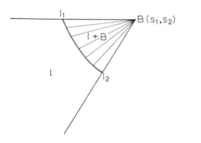

Fig. 114. Tie lines intersecting at corner B
of the concentration triangle.

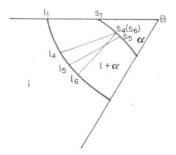

Fig. 115. Impossibility of intersection
of tie lines.

(4) Konovalov's Rule. The isothermal section at T_3 (Fig. 113) is taken between T_B and T_A. Isothermal sections at T_A and between T_A and T_C are given in Figs. 116a and b respectively. It will be noticed that the liquidus and solidus lines meet at corner A in Fig. 116a because this section represents the temperature corresponding to the melting point of A. Comparison of Figs. 113, 116a and 116b shows that the tie lines rotate in a counterclockwise direction as the

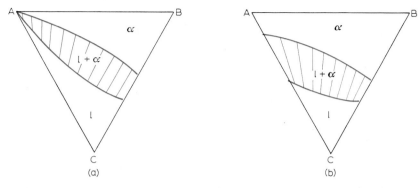

Fig. 116. Isothermal sections through Fig. 111 at (a) T_A, and (b) between T_A and T_C.

temperature falls and this rotation coincides with the direction from the highest melting point component to the lowest—from B to A and then to C.

This rotation is reflected in Konovalov's Rule, which states that the solid is always richer than the melt with which it is in equilibrium in that component which raises the melting point when added to the system. In binary systems (Fig. 27b)

$$X_A^s > X_A^l.$$

As

$$X_A^s + X_B^s = X_A^l + X_B^l = 1$$

then

$$\frac{X_A^s}{X_A^s + X_B^s} > \frac{X_A^l}{X_A^l + X_B^l}$$

and

$$\frac{X_A^s}{X_A^s + X_B^s - X_A^s} > \frac{X_A^l}{X_A^l + X_B^l - X_A^l} \text{ since } X_A^s > X_A^l.$$

Therefore

$$\frac{X_A^s}{X_B^s} > \frac{X_A^l}{X_B^l}.$$

In this form Konovalov's Rule can be applied to ternary systems to indicate the direction of tie lines.

Figure 117 reproduces Fig. 113 but only one tie line, ls, is shown. The lines from B through s and l intersect the side AC of the triangle at points s^1 and l^1 respectively. Then,

$$\frac{X_A^l}{X_C^l} = \frac{l^1 C}{l^1 A} \quad \text{and} \quad \frac{X_A^s}{X_C^s} = \frac{s^1 C}{s^1 A},$$

it being realised that the melting point of A is higher than that of C. As $s^1 C / s^1 A > l^1 C / l^1 A$ then $X_A^s / X_C^s > X_A^l / X_C^l$, i.e. the relative positions of points l and s are in agreement with Konovalov's Rule. Similarly, by drawing lines from A and C through points l and s it can be proved that $X_B^s / X_C^s > X_B^l / X_C^l$ and $X_B^s / X_A^s > X_B^l / X_A^l$. In this way it has been proved that Konovalov's Rule applies to each pair of components with regard to the relative positions of points l and s. The tie line ls is rotated anticlockwise by an angle θ relative to the line Bx^1. The latter line is obtained by joining B to the composition of the alloy X under consideration and extending this until it intersects the side AC. It is obvious that, for $\theta = 0$, then $X_A^s / X_C^s = X_A^l / X_C^l$ in contradiction to Konovalov's Rule. This also implies that tie lines when produced do not intersect the corner of the concentration triangle. It can be easily proved that Konovalov's Rule is contradicted if the point s lies to the right of curve Bx^1. In Fig. 117 the tie line ls is rotated anticlockwise to the line Bx^1 and again it will be noticed that this direction of rotation coincides with the direction of decreasing melting points of the components.

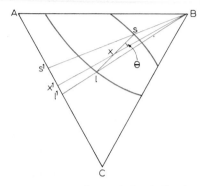

Fig. 117. Konovalov's rule for tie line ls at
a temperature T_3.

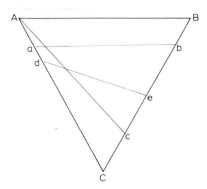

Fig. 118. Types of vertical section through
ternary space models.

(5) Experimental methods for determining tie lines across two-phase regions are summarised by Hume-Rothery, Christian and Pearson*.

Returning to the isothermal sections, it remains to mention that the section corresponding to the melting point of component C will show the existence of α solid solution throughout the concentration triangle ABC. This condition will also hold for lower temperatures since no solid state reactions were postulated in this system.

Vertical sections

Vertical sections** through ternary space diagrams are determined experimentally by studying the reactions of a series of alloys whose compositions lie on a straight line,

(a) parallel to one or other of the binary systems, such as line ab in Fig. 118,

(b) joining one corner of the triangle to any point on the opposite binary, such as line Ac in Fig. 118. Occasionally, sections are taken intermediate to the above two—line de.

Sections such as ab and Ac through Fig. 111 yield diagrams of the type shown in Fig. 119a and b. Both sections are similar to binary diagrams in that the same phase fields—l, $l+\alpha$, and α—appear in each, but they differ from the binary diagram in that the two curved lines only meet when the section is taken from a corner of the triangle (Fig. 119b). The upper line in each section is the liquidus line as it is generated by the intersection of the vertical plane of the section with the liquidus surface; the lower line is the solidus line generated by intersection with the solidus surface.

A more fundamental difference between vertical sections in ternary systems and the equivalent binary system is that, in general, the liquidus and solidus lines in vertical sections are not conjugate lines as they are in the binary phase diagram. In vertical sections the horizontal lines

---

* W. Hume-Rothery, J. W. Christian and W. B. Pearson, *Metallurgical Equilibrium Diagrams*, Institute of Physics, London, 1952, p. 285.
** The term *isopleth* is sometimes used to denote a vertical section. There seems no reason to adopt this additional nomenclature.

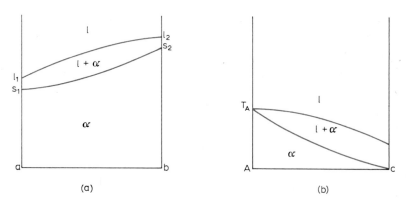

Fig. 119. Vertical sections through Fig. 111. (a) Section such as *ab* of Fig. 118; (b) section such as *Ac* of Fig. 118.

joining the compositions of conjugate liquid and solid phases are not tie lines. This important fact is illustrated by reference to Fig. 120, which shows the vertical section *ab* imposed on the three-dimensional space diagram of Fig. 111. The plane of the vertical section *ab* intersects the binary liquidus of the system AC at point l_1. The conjugate point to l_1 for the solid phase is point s_4 which is outside the plane of section *ab*. Similarly, the conjugate point for the liquid phase in equilibrium with the solid phase of composition s_1 on the binary solidus line $T_A s_4 s_1 T_C$ lies to the right of the plane of the vertical section. Considering point l_3 on the liquidus line

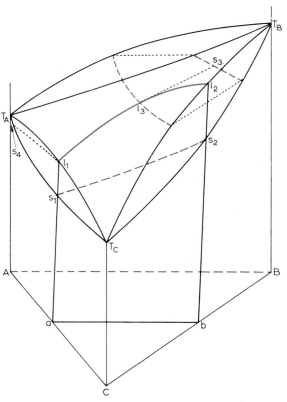

Fig. 120. The vertical section *ab* in relation to the space model of Fig. 111.

$l_1 l_3 l_2$ of section ab (Figs. 119a and 120) the conjugate solid phase is located at point s_3 on the solidus surface of the ternary diagram (the isothermal section at a temperature corresponding to the conjugate points l_3 and s_3 is shown dotted in Fig. 120). It is obvious that s_3 lies behind the plane of the section ab and that the tie line $l_3 s_3$ meets the vertical section at an angle.

Thus in general it is not possible to draw horizontal tie lines across two-phase regions in vertical sections to indicate the true compositions of the co-existing phases at a given temperature. The only time when this is permissible is when a section occurs in the diagram which is truly a binary section. This case can arise if a congruently-melting compound occurs in one of the binary systems. The section from the third component to the compound can then be a binary section—usually called a "pseudobinary section"—and tie lines can be drawn on such a vertical section to represent compositions of conjugate phases.

Although the compositions of co-existing phases cannot be read from vertical sections through a ternary system one can read the liquidus and solidus temperatures for any alloy represented in the section. Vertical sections are also extremely useful for demonstrating the effect of the addition of a third component to a binary system, especially when one component exhibits allotropy.

Polythermal projections

Polythermal projections are projections of the reactions occurring in the system at all temperatures on to the concentration triangle ABC. They are in effect a two-dimensional representation of the three-dimensional space diagram. Such projections are unnecessary in two-phase systems but they are useful in complex systems where it is difficult to draw an adequately clear space diagram.

Equilibrium freezing of alloys

All ternary alloys in the system under consideration behave in a similar manner on cooling from the liquid state. It has already been noted that the sequence of phase fields in this ternary system—l, $l+\alpha$, α—is the same as in a binary system showing a continuous series of solid solutions. It would be expected that the transformations in the ternary system on cooling or heating would be analogous to the transformations in the binary alloys. Considering an alloy X (Fig. 121) the vertical line $X s_3(l)$ represents the alloy composition at all temperatures. On cooling this alloy from the liquid state a temperature is reached corresponding to the liquidus surface. The liquidus surface in this case is reached at point l and, at this moment, the liquid becomes saturated with respect to the α solid solution. The initial composition of the α phase is represented by point s and the horizontal line joining the conjugate liquid and solid compositions is therefore a tie line.

With further cooling the liquid composition moves in a direction away from the original tie line ls as the solid phase precipitated is enriched with respect to component B and the liquid phase impoverished with respect to this component. The liquid composition moves along a curved path $l l_1 l_2 l_3$ on the liquidus surface during the solidification process, and the solid composition also moves along a similar curved path $s s_1 s_2 s_3$ on the solidus surface, each point on the liquidus surface being connected through the average alloy composition to a conjugate phase on the solidus surface. In this way the tie lines ls, $l_1 s_1$, $l_2 s_2$, $l_3 s_3$ represent the equilibrium compositions at decreasing temperatures during solidification. At a temperature corresponding to the tie line $l_3 s_3$ the composition of the solid phase has reached the average composition of the alloy X. At this temperature therefore the last drops of liquid are used up and the alloy becomes completely homogeneous α phase. Application of the lever rule to the series of tie lines ls, l_1s_1, … allows the proportion of liquid and solid phases to be calculated at each temperature during solidification. The composition of the final solid phase is therefore represented by s_3, which is vertically under point l in Fig. 121 since l represents the average composition of the alloy at a temperature corresponding to the initial tie line ls as well as the composition of the liquid which becomes saturated with and begins to precipitate α solid solution. In Fig. 121 this point has been denoted $s_3(l)$ to stress this fact. The manner in which the liquid and solid compositions change during solidification are precisely analogous to the mechanism operative in binary systems of this type.

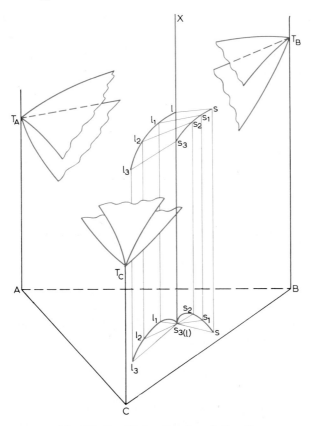

Fig. 121. Equilibrium freezing of alloy X.

The tie lines ls, l_1s_1, …, l_3s_3 generate a skew surface since they rotate round the vertical axis (skew axis) $Xs_3(l)$ as the temperature falls. This rotation of tie lines has been considered on p. 128 and becomes apparent from a glance at Figs. 122a–d*. The end of the tie line representing the liquid composition can be visualised as rotating towards the observer with fall in temperature whilst the opposite end of the tie line, representing the solid composition, rotates away from the observer. Rotation is round the axis of the line $Xs_3(l)$ since all tie lines connecting conjugate liquid and solid phases pass through the average composition of the alloy at all temperatures.

---

* After F. N. RHINES, *Phase Diagrams in Metallurgy*, McGraw-Hill, New York, 1956, p. 121.

The projection of the curves $ll_1l_2l_3$ and $ss_1s_2s_3$ on the concentration triangle gives a curve $ss_1s_2s_3(l)\ l_1l_2l_3$ with a cusp at point $s_3(l)$. At a temperature corresponding to the beginning of solidification, liquid l is in equilibrium with α phase of composition s. With the precipitation of a minute amount of α phase the composition of the liquid changes in the direction of the tie line ls away from component B, *i.e.* the original tie line ls is a tangent to the curve $ll_1l_2l_3$ of liquid compositions. In a similar way the tie line l_3s_3 is tangential to the curve $s_3s_2s_1s$ of solid compositions since on heating alloy X the first drop of liquid formed will change the composition of the solid phase in the direction of the tie line l_3s_3 towards the component B. Since all tie lines pass through the average alloy composition, the point $s_3(l)$ is common to all tie lines.

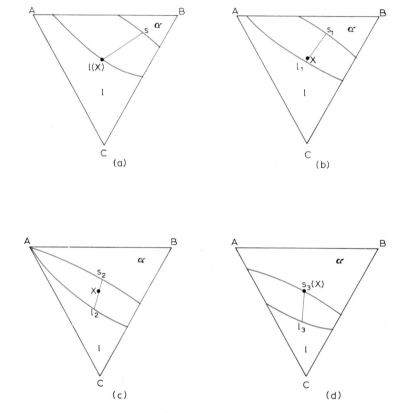

Fig. 122. Rotation of the tie lines ls, l_1s_1, l_2s_2 and l_3s_3 as alloy X solidifies. (After F. N. RHINES; courtesy McGraw-Hill.)

The polythermal projection on the concentration triangle again indicates the direction of rotation of the tie lines with fall in temperature.

Throughout the solidification two phases are present and the system is therefore bivariant. Referring to p. 118, two variables can be selected and then the system is exactly defined.

(1) *Selection of T and* X_A^l. If we select point l_1 on the isothermal section (Fig. 122b), the composition of the solid phase is given by the tie line l_1s_1 as point s_1.

(2) *Selection of* X_A^l *and* X_C^l (Fig. 123a). If vertical sections are drawn at a constant % A and a constant % C, parallel to edges BC and AB respectively, they will intersect along a vertical line l_1x. The vertical line meets the liquidus surface at point l_1. This point lies on the isotherm at temperature T, and the tie line on the isotherm from l_1 to s_1 defines the composition of the

solid. As the temperature and compositions of both the solid and liquid phases have now been determined the system is exactly defined.

(3) *Selection of X_C^l and X_C^s* (Fig. 123b). Two vertical sections are drawn parallel to edge AB at different values of X_C to represent the concentration of C in the liquid and solid phases. The trace of these planes on the ternary liquidus and solidus surfaces is given by curves ll_1l_2 and ss_1s_2. A tie line can only be drawn at a temperature T for points l_1 on the liquidus and s_1 on the solidus. Thereby the temperature and compositions of the two phases are defined.

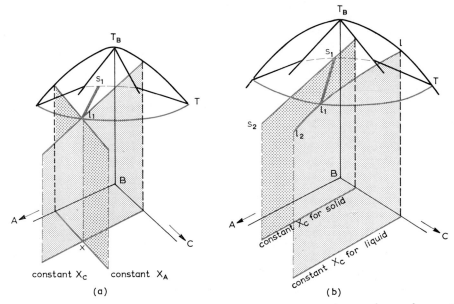

Fig. 123. Selection of two variables to define the system exactly, (a) X_A^l and X_C^l, (b) X_C^l and X_C^s.

Using the concept of free energy surfaces the equilibria in the system can be illustrated by Figs. 124a–c. As the temperature falls, the free energy surface of the liquid phase rises relative to that of the solid phase until, at T_C, the only contact of the two surfaces is at component C. Below T_C the liquid free energy surface lies completely above that of the solid phase and all alloys are stable in the solid state only.

An interesting example of this type of system is to be found in the quasi-ternary section UFe_2–UCo_2–UMn_2 of the Co–Fe–Mn–U quaternary system*. The intermediate phases AB_2 are acting as components of the quasi-ternary system.

The process of solidification described in this section assumes a sufficiently slow cooling to allow equilibrium conditions to be maintained at all temperatures. In particular, it is assumed that sufficient time is available for diffusion to occur in the solid state so that the composition of the α phase always reaches the equilibrium value at each temperature. Non-equilibrium cooling leads to coring of the α phase and these effects can be eliminated in the same way as for the corresponding binary alloys.

* G. Petzow, S. Steeb and G. Kiessler, Z. *Metallk.*, **54** (1963) 473.

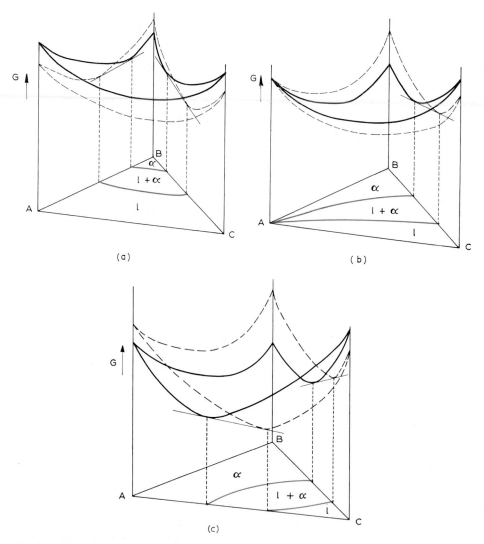

Fig. 124. Illustration of the equilibria in Fig. 111 by means of free energy surfaces for the liquid and α phases at various temperatures, (a) between T_B and T_A, (b) at T_A, and (c) between T_A and T_C.
– – – – –, liquid free energy surface; ————, α free energy surface.

8.4.2. *Variants of the phase diagram*

The only system worthy of consideration is that in which there is a minimum in two of the binary systems and a maximum in the third. This system, which is depicted in Fig. 125, has a saddle point* on the liquidus and solidus surfaces. In the direction from T_A to e the liquidus goes through a minimum point S; in the direction from point d to f the liquidus surface goes through a maximum point which is the same point S. At point S the liquidus and solidus surfaces of the ternary system meet; they also meet at the binary maximum e and the binary minima d and f as well as at T_A, T_B, and T_C. Figure 126 illustrates the projection of liquidus isothermals on

---

* Or a hyperbolic point since the liquidus surface has the shape of a hyperbolic paraboloid—as in a horse saddle.

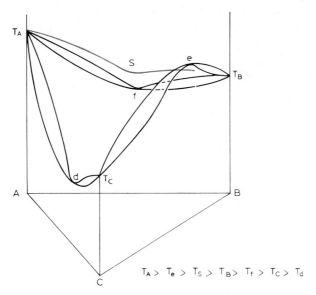

$$T_A > T_e > T_S > T_B > T_f > T_C > T_d$$

Fig. 125. Ternary two-phase equilibrium with a saddle point.

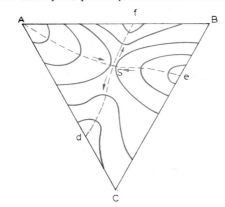

Fig. 126. Projection of the liquidus isotherms of Fig. 125 on the concentration triangle *ABC*.

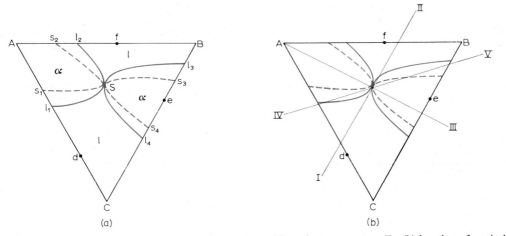

(a)   (b)

Fig. 127. (a) Isothermal section through Fig. 125 at the saddle point temperature, T_S; (b) location of vertical sections I–II, *A*–III and IV–V.

the concentration triangle. The dotted line ASe is the locus of the maximum points on the liquidus surface; the dotted line dSf is the locus of the minimum points on the liquidus surface. The point of intersection of these two lines, S, is called the saddle point.

Isothermal sections at temperatures above and below the saddle point are easily constructed and so attention will be directed to the section at the saddle point temperature. This section is shown in Fig. 127a, the continuous lines l_1Sl_2 and l_4Sl_3 being the liquidus isotherm and the dotted lines s_1Ss_2 and s_4Ss_3 are solidus isotherms. The areas As_2Ss_1A and $s_4Ss_3s_4$ are α solid solution region; the areas $s_1l_1Sl_2s_2Ss_1$ and $s_4l_4Sl_3s_3Ss_4$ are $l+\alpha$ regions, and the areas Bl_2Sl_3B and Cl_1Sl_4C are liquid regions.

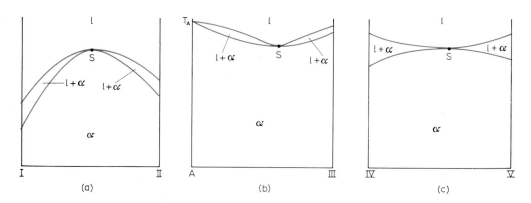

Fig. 128. Vertical sections through Fig. 125. (a) I–II; (b) A–III and (c) IV–V.

The vertical sections to be considered are shown in Fig. 127b. Section I–II (Fig. 128a) exhibits a maximum in both the liquidus and solidus surfaces. Section A–III (Fig. 128b) is characterised by a minimum in the liquidus and solidus. On the other hand section IV–V (Fig. 128c) possesses a minimum in the liquidus and a maximum in the solidus. All alloys in the system solidify as homogeneous α solid solution.

8.4.3. Two-phase equilibrium between solid or liquid solutions: $\alpha_1 \rightleftharpoons \alpha_2$ or $l_1 \rightleftharpoons l_2$

When a miscibility gap occurs in one or more binary systems a two-phase equilibrium can occur in the ternary system. The binary miscibility gap can occur in either the solid or liquid state with an equilibrium of the type $\alpha_1 \rightleftharpoons \alpha_2$ or $l_1 \rightleftharpoons l_2$. The ternary equilibrium can be treated on either basis—we will assume a solid miscibility gap of the type shown in Fig. 31 in one binary system only.

A ternary system with only a binary critical point
It will be assumed that the binary system AB contains a miscibility gap with a critical point c_1 at a temperature T_c. In the ternary system ABC the miscibility gap emanating from the binary AB must be closed, since the binary systems AC and BC are completely miscible in the solid state. We will consider initially the room-temperature isothermal section of the ternary (Fig. 129). The binary tie line $\alpha_1\alpha_2$ joins the compositions of the co-existing binary phases at room-temperature. Since the binary miscibility gap is closed in the ternary system on addition of C to the binary AB the compositions of the co-existing phases approach each other until,

at point c, they become identical. Point c is the critical point or the plait point and the curve $\alpha_1 c \alpha_2$ the solubility isotherm at the temperature in question. The tie lines $\alpha_1^1 \alpha_2^1$, $\alpha_1^2 \alpha_2^2$, ..., $\alpha_1^5 \alpha_2^5$ represent the compositions of the co-existing phases with increasing additions of C.

It will be noticed that the tie lines are not parallel to the binary tie line $\alpha_1 \alpha_2$. This is so because on the addition of C to a heterogeneous mixture of A and B, the component C does not distribute itself equally between the two liquids but in a ratio corresponding to the distribution coefficient of C between A and B. It follows that the maximum on the curve $\alpha_1 c \alpha_2$ does not necessarily coincide with the critical point, but will usually be displaced to one side of it. In Fig. 129 the maximum point on the solubility curve is point α_1^5, to the left of the critical point c.

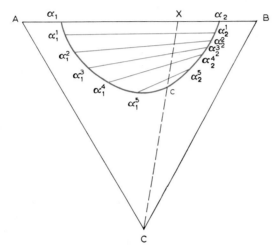

Fig. 129. Isothermal section at room temperature of a ternary system with a closed miscibility gap.

The solubility curve $\alpha_1 c \alpha_2$ separates the heterogeneous from the homogeneous regions of the system. Inside the parabolic area $\alpha_1 c \alpha_2$ a two-phase $\alpha_1 + \alpha_2$ equilibrium exists; outside this area the system is single-phase $\alpha_{1(2)}$. In the homogeneous region the Phase Rule indicates that the system is trivariant. In the two-phase region the system is bivariant. Thus if we choose a temperature and fix the concentration of component C in one of the co-existing phases, then the equilibrium is established. Conditions are different at the critical point for here we place a limitation on the system by requiring the compositions of the co-existing solid phases to be identical. In this way we have reduced the variables by one, and the system at the critical point will therefore be monovariant. It is then only necessary to specify the temperature, for instance, to define the equilibrium. Thus the composition of the critical point depends solely on the temperature.

The effect of temperature. The binary phase diagram (Fig. 31) indicates a miscibility gap which narrows with increase of temperature. In the ternary diagram this behaviour is reflected by a shrinkage of the miscibility gap (Fig. 129) with increase of temperature. The resulting phase diagram is illustrated in Fig. 130. A critical curve* $c c_1$ is generated from the individual isothermal critical points such as that represented in Fig. 129. The critical curve must have an end point at the binary critical point c_1 since the curve represents the temperature–composition

---

* Curve $c c_1$ is also called a plait point curve.

variation of ternary critical points (*i.e.* points indicating the compositions at which two solids become identical in state). This is also true of the binary critical point c_1.

It will be noted that the maximum point on the solubility curves generates a maximum curve $c_1 m$ distinct from the critical curve $c_1 c$. The critical curve and the maximum curve meet at the binary critical point c_1 since the maximum on the binary diagram must coincide with the binary critical point. Although mention has been made of individual critical points on the ternary isotherms there is no true ternary critical point in the system under discussion. The surface generated by the shrinking of the room-temperature–solubility curve is called the solubility surface, $\alpha_1 c \alpha_2 c_1 c$.

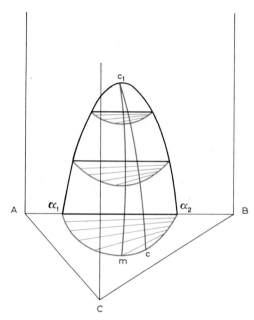

Fig. 130. Space model of a ternary system with a closed miscibility gap associated with a binary critical point c_1.

The transformations occurring in an alloy whose ordinate cuts the critical curve will be considered. In Fig. 131a the ordinate of alloy X cuts the critical curve $c_1 c$ at point a_2. At a temperature T_1 alloy X will be single-phase since it is in the $\alpha_{1(2)}$ phase region. On cooling to T_2 the alloy reaches the solubility surface at point a_2. With a slight drop in temperature below T_2 the alloy will break up into two solids α_1 and α_2, thus establishing a two-phase equilibrium. On further cooling the points representing the compositions of these conjugate solids will each trace a curve over the solubility surface. Solid α_1 traces the curve $a_2 a_3^1 a_4^1$ and solid α_2 the curve $a_2 a_3^{11} a_4^{11}$. The course of the curves is defined by the relative position of the tie lines which skew round towards the side AB as the temperature decreases. At T_2 the tie line degenerates to a point a_2. The initial path of both curves $a_2 a_3^1 a_4^1$ and $a_2 a_3^{11} a_4^{11}$ (Fig. 131b) is along a line which is tangential to the solubility curve $\alpha_1^1 a_2 \alpha_2^1$ at a_2. Projecting the path of the two solids on to the concentration triangle the curve $a_4^1 a_3^1 a_2 a_3^{11} a_4^{11}$ is obtained. It is a continuous curve, with a_2 as a point of inflection; the part $a_4^1 a_3^1 a_2$ is convex to the critical point c and the part $a_2 a_3^{11} a_4^{11}$ is convex to the binary side AB. The shape of the curve $a_4^1 a_3^1 a_2 a_3^{11} a_4^{11}$ is a direct consequence of the skewing of the tie lines with fall in temperature. The skew surface $a_2 a_3^1 a_4^1 X a_4^{11} a_3^{11} a_2$ is formed by rotation of the tie lines about the vertical axis $X a_2$.

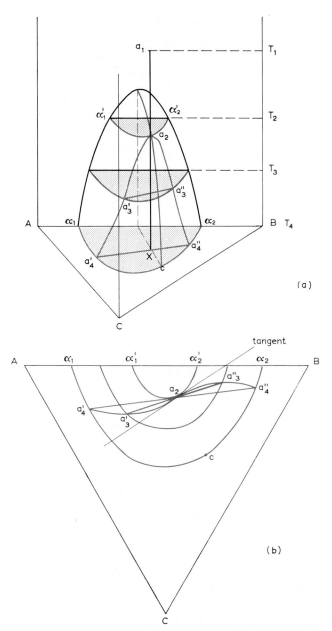

Fig. 131. (a) Transformations in alloy X on cooling from the $\alpha_{1(2)}$ phase region; (b) changes in composition of the co-existing α_1 and α_2 phases illustrated by projection of Fig. 131a on concentration triangle ABC.

The greater the displacement of the critical points from the maximum points on the solubility curves, the greater will be the rotation of the tie lines with fall in temperature, and therefore the greater the skew on the surface representing the separation of the two solids.

This type of ternary system, but with a miscibility gap in the liquid state extending from one of the binary systems was the first ternary system to be investigated in any detail*. The system Au–Pd–Pt** is an example of a system with a solid state miscibility gap.

———————

*,** See overleaf.

A ternary system with a binary and a ternary critical point

The second case of a miscibility gap in only one of the binary systems leads to a ternary critical point, as in the Au–Cu–Ni system***. This occurs if the ternary miscibility gap extends to a higher temperature than the binary miscibility gap (Fig. 132). The addition of C to the binary critical mixture c_1 raises the temperature of stability of the two-phase field. A ternary critical point c_3 is formed with a critical curve $c_1 c_3 c$. The ternary critical point, like its binary counterpart, is also the maximum point on the solubility surface.

Isotherms are given in Fig. 132. Of special interest is the isotherm corresponding to the temperature of the binary critical point c_1 (Fig. 133a). This isotherm is a continuous closed

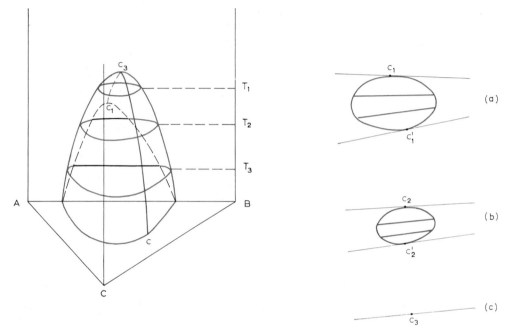

Fig. 132. Space model of a ternary system with a closed miscibility gap associated with a ternary critical point c_3.

Fig. 133. Isotherms through Fig. 132 at (a) the binary critical point temperature, (b) a temperature between c_1 and c_3, (c) the ternary critical point temperature.

curve with two critical points, c_1 and c_1^1. The tangent to the binary critical point c_1 can be considered as giving the direction of the tie line at the binary critical temperature. It coincides with the direction of the side AB. In reality, the tie line degenerates to a point at c_1 but the tangent to c_1 represents the hypothetical tie line. In a similar manner the tangent to c_1^1 represents the direction of the tie line in the ternary system. With increase of temperature the isotherm shrinks away from the side AB but still maintains two critical points (Fig. 133b). These points, c_2 and c_2^1, move closer together with increase in temperature until they eventually converge at the ternary critical point c_3 (Fig. 133c). The tangents acquire the same direction and, as before, the direction of the tangent is that of the hypothetical degenerate tie line corresponding to the ternary critical point.

* See footnote, p. 120. It is of interest to note that Stokes also introduced the terms "conjugate" and "tie line" to express the equilibrium of two phases at a particular temperature.
** E. RAUB AND G. WÖRWAG, *Z. Metallk.*, **46** (1955) 513.
*** E. RAUB AND A. ENGEL, *Z. Metallk.*, **38** (1947) 11.

CHAPTER 9

Ternary Phase Diagrams. Three-Phase Equilibrium

9.1. PROPERTIES OF THREE-PHASE TIE TRIANGLES

In a binary system the equilibrium of three phases is invariant and as such is represented by a horizontal line (isotherm) in the phase diagram. The distinguishing feature of ternary three-phase equilibrium is that, being monovariant, it occurs over a temperature range. At any one temperature the composition of the three conjugate phases is fixed (p. 119). As the compositions of the three phases generally vary with the temperature, we can represent the three-phase equilibrium by three curves: $\alpha\alpha_1$, $\beta\beta_1$ and ll_1 (Fig. 134). At a temperature T the three phases α, β and l are in equilibrium; lines $\alpha\beta$, αl and βl are tie lines and $\alpha\beta l$ is a tie triangle. From T to T_1 the state space for $\alpha + \beta + l$ equilibrium is constructed from a series of tie triangles. It follows from this that the surfaces $\alpha\beta\beta_1\alpha_1$, $\alpha ll_1\alpha_1$ and $\beta ll_1\beta_1$ are ruled surfaces. They are not planar surfaces since in general the surfaces skew with fall in temperature.

Certain characteristics of three-phase triangles can be summarised:

(1) As already noted, all three-phase regions appear as triangles in isothermal sections. All alloys within the three-phase triangle consist of the phases whose compositions are represented by the vertices of the triangle. In general, the composition of these phases changes with variation in temperature.

(2) Each three-phase triangle on an isothermal section is associated with three two-phase regions.

(3) Each three-phase triangle on an isothermal section makes contact at each vertex with a one-phase region.

(4) Each binary three-phase reaction is associated with an equivalent monovariant reaction in the ternary system. A binary eutectic reaction, for example, implies the presence of a three-phase equilibrium between a liquid and two solids in the ternary system.

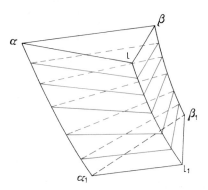

Fig. 134. Ternary three-phase equilibrium.

(5) In isothermal sections the boundaries between a one-phase and a two-phase region where they meet the three-phase triangle have a certain direction, which was originally stated* as follows:

When two equilibrium curves intersect then their metastable extensions in the neighbourhood of the point of intersection lie either inside or outside the corresponding three-phase triangle.

Vogel** has qualified this generalisation to the extent that the acute angle which the solubility isotherms make must point towards the three-phase triangle. In accordance with these proposi-

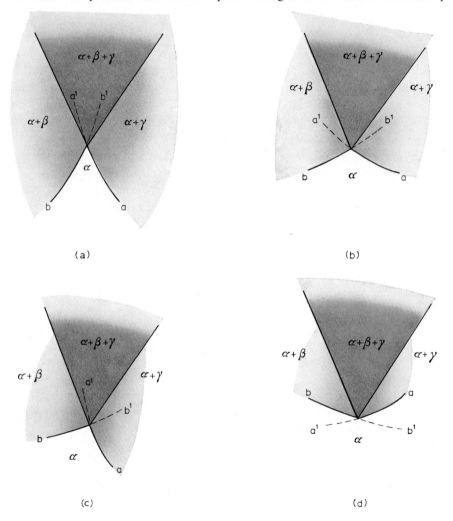

Fig. 135. Disposition of phase boundaries in isothermal sections of ternary systems containing three-phase regions. (a) and (b) are possible boundary positions; (c) and (d) are impossible boundary positions.

---

* F. A. H. SCHREINEMAKERS, in H. W. BAKHUIS ROOZEBOOM (ed.), *Die Heterogenen Gleichgewichte vom Standpunkte der Phasenlehre*, Vol. 3, Part 2, Braunschweig, 1913, p. 116.
** R. VOGEL, *Die Heterogenen Gleichgewichte*, Geest & Portig, Leipzig, 2nd. edn., 1959, p. 362.

tions the disposition of curves in Figs. 135a and b are possible; the cases represented by Figs. 135c and d are impossible. A thermodynamic treatment was published by Schreinemakers and by Lipson and Wilson*.

9.2. THREE-PHASE EQUILIBRIUM
PRODUCED BY THE COALESCENCE OF MISCIBILITY GAPS

A three-phase equilibrium can be produced by the coalescence of two two-phase regions (Figs. 136 and 138). Referring to Fig. 136, the two-phase regions move closer together as the temperature falls until, at T_2, contact is made at point k. At temperatures immediately below T_2 a three-phase equilibrium involving phases of composition a, b and c is formed. The tie line kd

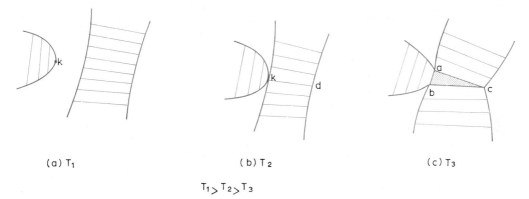

(a) T_1 (b) T_2 (c) T_3

$T_1 > T_2 > T_3$

Fig. 136. Production of a ternary three-phase equilibrium by the coalescence of two two-phase regions.

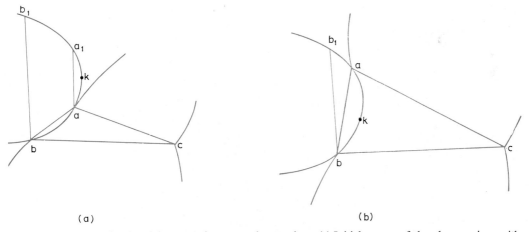

(a)   (b)

Fig. 137. Conditions for the coalescence of two two-phase regions. (a) Initial contact of the phase regions with point k outside curve ab; (b) initial contact with point k on curve ab.

at T_2 can be thought of as a degenerate tie triangle since point k is a degenerate tie line (cf. tie line ab which degenerates to the point k in going from T_3 to T_2). It has been assumed that the two-phase regions make contact at the critical point k on the left-hand two-phase region (Figs. 136a and b). This can be proved by assuming that contact does not take place at k.

---

* H. LIPSON AND A. J. C. WILSON, *J. Iron Steel Inst.* (*London*), **142** (1940) 107.

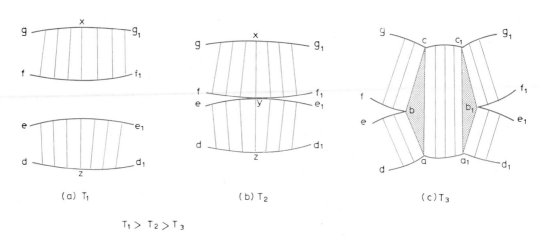

Fig. 138. Alternative method to Fig. 136 for the production of a ternary three-phase equilibrium by the coalescence of two two-phase regions.

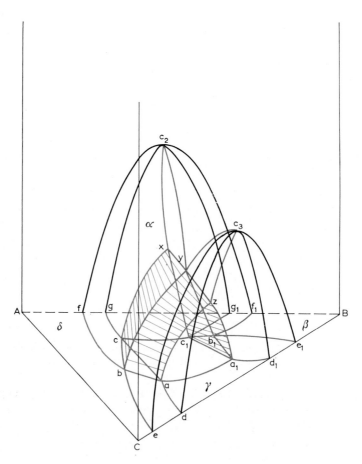

Fig. 139. Space model of a ternary system corresponding to Fig. 138.

If this were the case, the three-phase equilibrium at T_3 would be as represented in Fig. 137a. But phase a is in equilibrium with phase a_1 and phase b with phase b_1; we now have a five-phase equilibrium between a, b, c, a_1 and b_1. This is impossible. We conclude that contact must take place such that point k lies between a and b. There are two possibilities. If we assume that phases a and b are not on the same tie line (Fig. 137b), we would produce a four-phase equilibrium between phases of composition a, b, c and b_1. We should also have the impossible condition of two tie lines, ab and b_1b, intersecting at b. We are now forced to conclude that phases a and b lie on the same tie line and with fall in temperature these phases approach point k, which is the first point of contact with the second two-phase region.

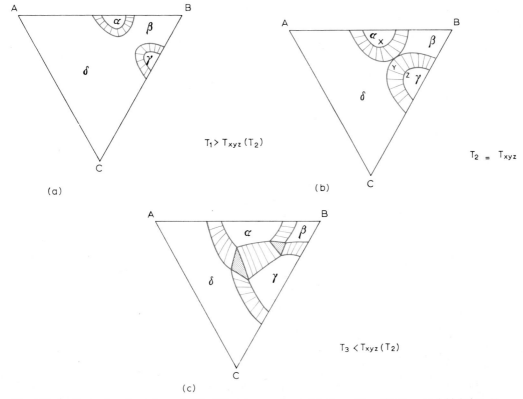

Fig. 140. Isothermal sections through Fig. 139 at temperatures (a) above T_{xyz}, (b) T_{xyz}, and (c) below T_{xyz}.

The second method for producing three-phase equilibrium is shown in Fig. 138. The two two-phase regions are separate at T_1; with fall in temperature they eventually touch at point y (Fig. 138b). At T_2 phases of composition x, y and z on the tie line xyz are in equilibrium. Further fall in temperature produces two three-phase tie triangles, abc and $a_1b_1c_1$, separated by a two-phase region acc_1a_1 (Fig. 138c). A ternary space diagram of a system involving this type of three-phase equilibrium is shown in Fig. 139. Three isothermal sections are given in Figs. 140a–c. It will be noticed that each corner of the three-phase triangles develops a curve as the temperature falls from $T_2(=T_{xyz})$. The curve developed by points a and a_1, b and b_1, c and c_1 meet at x, y and z (temperature T_2) respectively, so the number of continuous curves in the space diagram is only three: cxc_1, byb_1 and aza_1 (Fig. 139). In other words, the three mono-variant curves cxc_1, byb_1 and aza_1 all pass through a maximum at the temperature of the

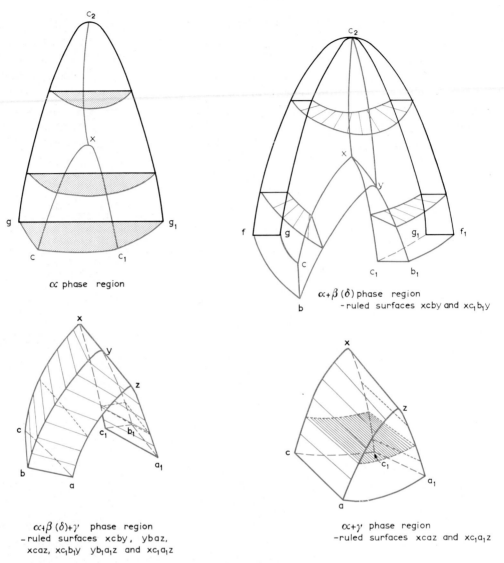

Fig. 141. Representation of the α, $\alpha+\beta(\delta)$, $\alpha+\beta(\delta)+\gamma$ and $\alpha+\gamma$ phase regions in Fig. 139.

degenerate tie triangle *xyz*. This point was originally derived by Gibbs* and stated in the following form by Schreinemakers**.

If the *n* phases of an *n*-component system have such a composition that a reaction is possible between them then at constant pressure the temperature is a maximum or a minimum.

Taking $n = 3$, it follows that the temperature is a maximum or a minimum if the three phases can be represented by points on a straight line.

To clarify the space diagram (Fig. 139) several of the phase regions are shown separately in Fig. 141. Isothermal sections are also included for further clarification. According to Raub and Wörwag*** the Au–Cu–Pd alloys have solid state equilibria of this type.

---

* *The Scientific Papers of J. Willard Gibbs*, Vol. 1, Dover Publications Inc., 1961, p. 99.

** F. A. H. SCHREINEMAKERS, in H. W. BAKHUIS ROOZEBOOM (ed.), *Die Heterogenen Gleichgewichte vom Standpunkte der Phasenlehre*, Vol. 3, Part 1, Braunschweig, 1911, p. 285.

*** E. RAUB AND G. WÖRWAG, *Z. Metallk.*, **46** (1955) 119. In the Au–Cu–Pd system curves c_2x and c_2y have a maximum critical point. Otherwise, the diagram is exactly as given in Fig. 139.

9.3. THREE-PHASE EQUILIBRIUM INVOLVING EUTECTIC REACTIONS

9.3.1. *A eutectic solubility gap in one binary system*

A eutectic solubility gap is assumed in one binary, and continuous series of solid solutions in the other two binaries. There are two types of ternary diagram, one with a minimum critical point on the ternary monovariant curve and one with a maximum critical point on that curve.

9.3.1.1. *A minimum critical point*

The space diagram

The space diagram is given in Fig. 142 and the projection of the system on to the concentration triangle in Fig. 143. The binary system AB is eutectic-type, $l_e \rightleftharpoons \alpha_a + \beta_b$, whereas binaries AC and BC show continuous series of solid solutions. Complete miscibility is assumed in the liquid state. With these conditions the eutectic solubility gap must be closed within the ternary system; it cannot stretch to either binary side AC or BC. A closed solubility loop is formed, stretching from a through c to b. It should be noted that the loop acb slopes downwards from the binary eutectic horizontal ab to c. The α solid solution along curve ac is in equilibrium with β along curve bc. Point c is therefore a critical point at which the ternary α and β phases become indistinguishable (designated $\alpha(\beta)$).

The eutectic curve which originates at point e on the binary AB must end somewhere in the ternary. As will become apparent later, this curve ee_1 ends at point e_1, where e_1 and c lie at the same temperature. Curve ee_1 appears on the liquidus surface as a trough which peters out

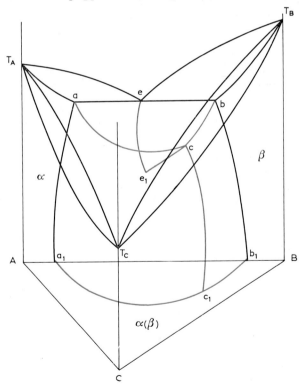

Fig. 142. Space model of a ternary system with a eutectic solubility gap in one binary system and a minimum critical point on the monovariant liquid curve.

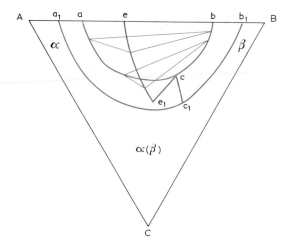

Fig. 143. Projection of Fig. 142 on the concentration triangle ABC.

at e_1 (Fig. 144). The liquidus isotherm for the temperature of point e_1 is shown as the continuous line $l_1e_1l_2$. It is convenient to use this line to describe the various parts of the liquidus surface but it should be remembered that line $l_1e_1l_2$ has no significance in the ternary diagram. The liquidus surface consists of three parts:

$T_Aee_1l_1T_A$ – the liquidus surface of α solid solution,

$T_Bee_1l_2T_B$ – the liquidus surface of β solid solution,

$l_1e_1l_2T_Cl_1$ – the liquidus surface of $\alpha(\beta)$ solid solution.

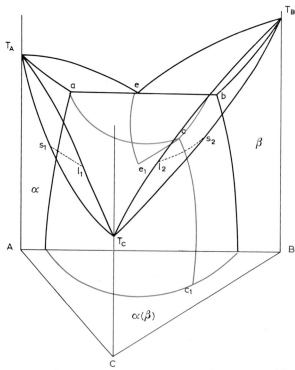

Fig. 144. Space model of Fig. 142 with isothermal section at the critical temperature superimposed.

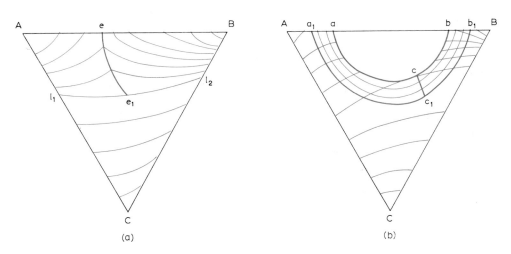

Fig. 145. Polythermal projection of (a) the liquidus surface, (b) the solidus surface and the solubility surface of Fig. 142.

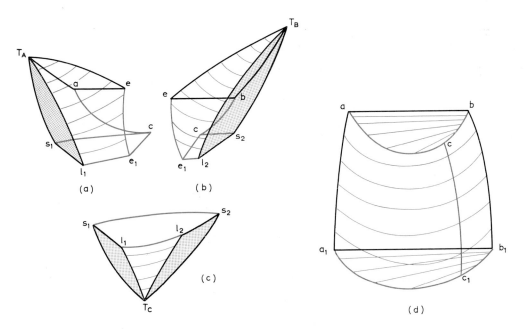

Fig. 146. The two-phase regions of Fig. 142.

(a) The $l+\alpha$ phase region. Boundary surfaces—liquidus $T_A l_1 e_1 e T_A$ (above); solidus $T_A acs_1 T_A$ and isothermal plane $l_1 e_1 cs_1 l_1$ (below); $l\alpha$ ruled surface aee_1ca (right); part $T_A l_1 s_1 T_A$ of $l+\alpha$ region of binary AC system (left); part $T_A ea T_A$ of $l+\alpha$ region of binary AB system (behind).

(b) The $l+\beta$ phase region. Boundary surfaces—liquidus $T_B l_2 e_1 e T_B$ (above); solidus $T_B bcs_2 T_B$ and isothermal plane $l_2 e_1 cs_2 l_2$ (below); $l\beta$ ruled surface bee_1cb (left); part $T_B l_2 s_2 T_B$ of $l+\beta$ region of BC system (right); part $T_B eb T_B$ of $l+\beta$ region of AB system (behind).

(c) The $l+\alpha(\beta)$ phase region. Boundary surfaces—liquidus $T_C l_1 e_1 l_2 T_C$ and isothermal plane $l_1 s_1 s_2 l_2 l_1$ (above); solidus $T_C s_1 cs_2 T_C$ (below); part $T_C s_1 l_1 T_C$ of $l+\alpha$ region of AC system (left); part $T_C s_2 l_2 T_C$ of $l+\beta$ region of BC system (right).

(d) The $\alpha+\beta$ phase region. Boundary surfaces—$\alpha\beta$ ruled surface $abca$ (above); room temperature isotherm $a_1 b_1 c_1 a_1$ (below); $\alpha+\beta$ solubility gap $abb_1 a_1 a$ of the AB system (behind); solubility surfaces $acc_1 a_1 a$ of β in α and $bcc_1 b_1 b$ of α in β (front).

Isotherm $l_1e_1l_2$ is convenient for defining the limits of existence of either α or β on the one hand, or $\alpha(\beta)$ solid solution on the other hand. A polythermal projection of the liquidus surface is given in Fig. 145a. In a similar manner the solidus surface can be visualised as divided into three parts by the solubility curve acb and the solidus isothermal s_1cs_2:

$T_As_1caT_A$ – the solidus surface of α solid solution,

$T_Bs_2cbT_B$ – the solidus surface of β solid solution,

$T_Cs_1cs_2T_C$ – the solidus surface of $\alpha(\beta)$ solid solution.

A polythermal projection of the solidus and solubility surfaces is given in Fig. 145b. The solubility surface $acbb_1c_1a_1a$ is divided into two by the curve of critical compositions, cc_1. The portion acc_1a_1a represents the solubility surface of β in α, and the portion bcc_1b_1b the solubility surface of α in β.

The two-phase regions $l+\alpha$, $l+\beta$, $l+\alpha(\beta)$ and $\alpha+\beta$ are shown separately in Fig. 146a–d.

The three-phase region

The three-phase $(l+\alpha+\beta)$ region extends from the temperature of the binary AB eutectic to the temperature of the tie line e_1c. Any isothermal section between these two temperatures will produce an $l+\alpha+\beta$ region with triangular shape. The vertices of the triangle are produced by intersection of the isothermal plane with corresponding points on curves ac, bc, and ee_1. Two such three-phase triangles are illustrated in Fig. 143. The liquid changes in composition along curve ee_1, whilst the co-existing α and β phases change composition along curves ac and bc respectively. The three-phase equilibrium ends when the compositions of the α and β phases become identical at the critical point c. At that temperature liquid of composition e_1 is in equilibrium with $\alpha(\beta)$ of composition c. It is obvious that line e_1c is a degenerate tie triangle. The $l+\alpha+\beta$ phase region is bounded by three ruled surfaces, as shown in Fig. 147 (*cf.* p. 143).

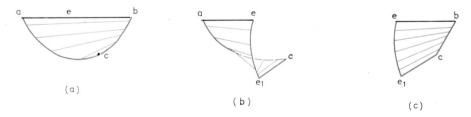

Fig. 147. The ruled surfaces bounding the three-phase $(l+\alpha+\beta)$ region in Fig. 142. (a) The $\alpha\beta$ ruled surface; (b) the $l\alpha$ ruled surface; (c) the $l\beta$ ruled surface.

Before considering the nature of the three-phase reaction in this system it is appropriate to list the ways in which three-phase regions terminate in ternary systems:

(1) *Termination at a degenerate tie triangle.* In the present system the $l+\alpha+\beta$ phase region ends at the degenerate tie triangle e_1c. Compare also the space diagram given in Fig. 139.

(2) *Termination at a reaction isotherm.* Referring to Fig. 148a, the three-phase region represented by the tie triangle $a^1e^1b^1$ ends at the reaction isotherm aeb. As will become apparent later, the reaction isotherm aeb is not necessarily a binary one but can also be a quasi-binary reaction isotherm in a ternary system or a ternary reaction isotherm such as a saddle point.

(3) *Termination at a four-phase plane.* The triangle $a^1e^1b^1$ ends on the four-phase plane abc (*cf.* p. 124). Point e lies on the plane abc (Fig. 148b).

(4) *Termination on the concentration triangle.* There are many examples of a three-phase

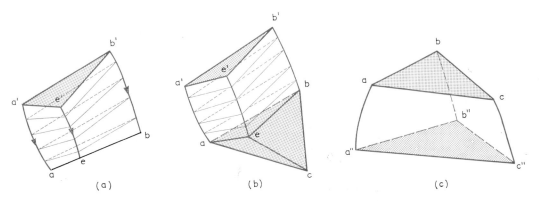

Fig. 148. Termination of three-phase regions, (a) at a reaction isotherm *aeb*, (b) at a four-phase plane, and (c) on the concentration triangle *ABC*.

triangle falling unchanged to the concentration triangle. The triangle *abc* (Fig. 148b) could fall from the four-phase plane to the concentration triangle as shown in Fig. 148c.

It would seem at first sight that the ternary three-phase equilibrium between liquid, α and β would be eutectic since it originates from the binary eutectic reaction. On this basis, liquid would deposit α and β until the liquid composition had reached point e_1 (Fig. 142) and the α and β had reached identical compositions at point *c*. Although the equilibrium is basically eutectic there are conditions under which it can be peritectic. The usual criterion for distinguishing between ternary three-phase eutectic and peritectic reactions is illustrated in Fig. 149. The tangent to the liquid curve at a particular temperature is extrapolated to meet the tie line connecting the α and β phases. If the extrapolated line intersected the $\alpha\beta$ tie line, the equilibrium was considered to be eutectic; if it met the $\alpha\beta$ tie line only when the latter was extrapolated, the equilibrium was considered to be peritectic. Similarly, a eutectoid reaction could be distinguished from a peritectoid and a monotectic from a syntectic.

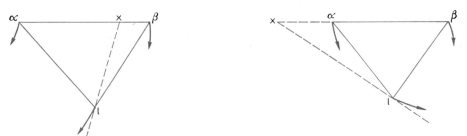

Fig. 149. Criteria for distinguishing eutectic and peritectic reactions in ternary three-phase equilibrium.

Ivanov* was the first to show that this criterion was incorrect. Later, Hillert** provided an alternative criterion. Both proved that, except for special cases, the relative amounts of the α, β and *l* phases are decisive in differentiating the type of ternary three-phase reaction undergone by an alloy on cooling. The average composition of the alloy then determines for a particular temperature whether the reaction will be eutectic or peritectic. Hillert's criterion will be described and then applied to the three-phase equilibrium in the present system.

---

* O. S. IVANOV, *Compt. Rend. Acad. Sci. U.R.S.S.*, **49** (1945) 349.

** M. HILLERT, *J. Iron Steel Inst.*, **189** (1958) 224. In the discussion of his paper (*ibid.*, **195** (1960) 201) Hillert gave a method for the graphical representation of his original mathematical treatment.

Hillert's criterion

An alloy within a three-phase triangle has a composition X_A, X_B given by:

$$X_A = \frac{m_\alpha X_A^\alpha + m_\beta X_A^\beta + m_l X_A^l}{m_\alpha + m_\beta + m_l}$$

$$X_B = \frac{m_\alpha X_B^\alpha + m_\beta X_B^\beta + m_l X_B^l}{m_\alpha + m_\beta + m_l}$$

A small change in temperature, dT, causes a small change in the composition and amounts of each phase, but not of the alloy itself,

i.e., $$\Delta X_A = 0, \quad \text{and} \quad \Delta m_\alpha + \Delta m_\beta + \Delta m_l = 0.$$

Therefore,

$$m_\alpha \Delta X_A^\alpha + X_A^\alpha \Delta m_\alpha + m_\beta \Delta X_A^\beta + X_A^\beta \Delta m_\beta + m_l \Delta X_A^l + X_A^l \Delta m_l = 0.$$

To simplify the calculation Hillert chooses the co-ordinate system X, V such that $X_A^\beta = X_A^l > X_A^\alpha$ (Fig. 150). Since

$$\Delta m_\beta + \Delta m_l = -\Delta m_\alpha$$

then,

$$\Delta m_\alpha (X_A^\beta - X_A^\alpha) = m_\alpha \Delta X_A^\alpha + m_\beta \Delta X_A^\beta + m_l \Delta X_A^l .$$

Since, by definition, $X_A^\beta > X_A^\alpha$, then Δm_α (the change in the amount of α on cooling dT) has the same sign as the expression $m_\alpha \Delta X_A^\alpha + m_\beta \Delta X_A^\beta + m_l \Delta X_A^l$.

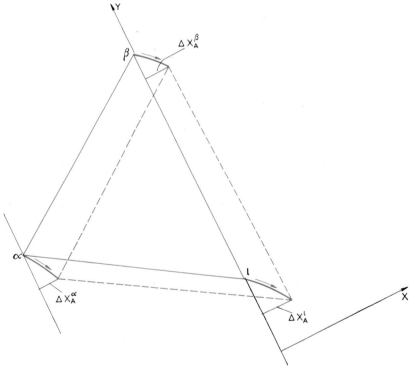

Fig. 150. Illustration of Hillert's criterion for distinguishing eutectic and peritectic reaction in ternary three-phase equilibrium; ————, equilibrium at T; — — — — — —, equilibrium at $T - dT$.

Similarly, if

$$X_A^\alpha = X_A^l > X_A^\beta$$

then

$$\Delta m_\beta(X_A^\alpha - X_A^\beta) = m_\alpha \Delta X_A^\alpha + m_\beta \Delta X_A^\beta + m_l \Delta X_A^l$$

and if

$$X_A^\alpha = X_A^\beta > X_A^l$$

then

$$\Delta m_l(X_A^\alpha - X_A^l) = m_\alpha \Delta X_A^\alpha + m_\beta \Delta X_A^\beta + m_l \Delta X_A^l .$$

Thus the change in the amount of the α, β and liquid phases has the same sign as the expression $m_\alpha \Delta X_A^\alpha + m_\beta \Delta X_A^\beta + m_l \Delta X_A^l$, when the co-ordinate system is suitably chosen. There are three cases:

(1) Δm_α is positive, Δm_β is positive and Δm_l is negative. The reaction is eutectic, $l \rightleftharpoons \alpha + \beta$.
(2) Δm_α is positive, Δm_β is negative and Δm_l is negative. The reaction is peritectic, $l + \beta \rightleftharpoons \alpha$.
(3) Δm_α is negative, Δm_β is positive and Δm_l is negative. The reaction is peritectic, $l + \alpha \rightleftharpoons \beta$.

The Hillert criterion indicates that the relative amounts of the α, β and liquid phases (the average alloy composition) are of importance in determining the type of reaction. This is so because the sign of Δm_α, for example, is dependent on the change in compositions of α, β and liquid. If ΔX_A^β is geometrically negative, the expression $m_\alpha \Delta X_A^\alpha + m_\beta \Delta X_A^\beta + m_l \Delta X_A^l$ will become negative as soon as the amount of β, m_β, is high enough. In a three-phase triangle, therefore, it is possible for some alloys to undergo eutectic reaction, whereas other alloys undergo peritectic reaction.

We will now return to a consideration of the three-phase reaction in Fig. 142. Seven isotherms

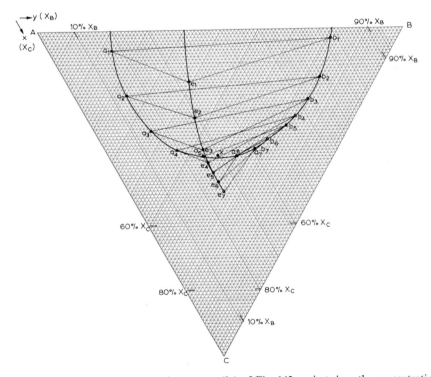

Fig. 151. Three-phase regions $a_1e_1b_1$, $a_2e_2b_2$, ..., $a_7e_7(b_7)$ of Fig. 142 projected on the concentration triangel.

through the three-phase region—$a_1e_1b_1$, $a_2e_2b_2$, ..., $a_7e_7(b_7)$—are shown on a projection of the three-phase region in Fig. 151. To determine whether the reaction is always a monovariant eutectic type, irrespective of alloy composition within the three-phase region, we apply Hillert's criterion to each pair of isotherms. Ideally, the isotherms should be spaced so closely that the difference in temperature between them is dT. The co-ordinates of the three-phase triangles are given in Table 2.

TABLE 2

	X_B,	X_C		X_B,	X_C		X_B,	X_C
e_1	33,	16	a_1	17,	6	b_1	78,	3
e_2	29,	27	a_2	14,	20	b_2	69,	15
e_3	26,	37	a_3	15,	31	b_3	62,	22
e_4	25.3,	41	a_4	19,	37	b_4	56,	27
e_5	25,	44	a_5	25,	39	b_5	52,	30
e_6	25,	47	a_6	34,	39	b_6	45,	34
e_7	25,	50	$a_7(b_7)$	40,	37			

Reactions in the triangle $a_1e_1b_1$. Consider triangles $a_1e_1b_1$ and $a_2e_2b_2$. Choose the X, Y axes such that $X_A^\beta = X_A^l > X_A^\alpha$, as in Fig. 150. ΔX_A^α, ΔX_A^β and ΔX_A^l are all positive, and therefore Δm_α is positive. This signifies that the amount of α increases on cooling from the temperature of isotherm $a_1e_1b_1$ to that of isotherm $a_2e_2b_2$. By changing the X, Y axes to give the conditions $X_A^\alpha = X_A^l > X_A^\beta$ it can be shown that Δm_β is positive. Although we assume that liquid is being consumed as the temperature falls, it can be shown that Δm_l is negative by choosing axes such that $X_A^\alpha = X_A^\beta > X_A^l$. Since Δm_α and Δm_β are positive and Δm_l is negative, the reaction undergone by any alloy within the triangle $a_1e_1b_1$ is eutectic-type: $l \rightleftharpoons \alpha + \beta$.

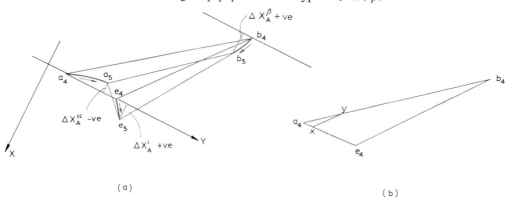

(a)   (b)

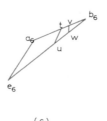

(c)

Fig. 152. The tie triangle $a_4e_4b_4$, (a) application of Hillert's criterion, (b) reaction equilibria. The tie triangle $a_6e_6b_6$, (c) reaction equilibria.

By examining triangles $a_2e_2b_2$–$a_3e_3b_3$ and $a_3e_3b_3$–$a_4e_4b_4$ it can also be demonstrated that the three-phase reaction is purely eutectic in character for triangles $a_2e_2b_2$ and $a_3e_3b_3$. Conditions are different for triangle $a_4e_4b_4$.

Reactions in the trangle $a_4e_4b_4$. Consider triangles $a_4e_4b_4$ and $a_5e_5b_5$. Again Δm_α is positive and Δm_l negative, but Δm_β can be either positive or negative for alloys in triangle $a_4e_4b_4$. This can be verified by choosing axes such that $X_A^\alpha = X_A^l > X_A^\beta$ (Fig. 152a). ΔX_A^α is negative, ΔX_A^β and ΔX_A^l are positive. Thus since Δm_β has the same sign as $m_\alpha \Delta X_A^\alpha + m_\beta \Delta X_A^\beta + m_l \Delta X_A^l$, Δm_β can be either positive or negative depending on the value of the negative term $m_\alpha \Delta X_A^\alpha$ in comparison with the positive term $m_\beta \Delta X_A^\beta + m_l \Delta X_A^l$. Obviously, if m_α is relatively large compared to m_β and m_l (if an alloy has a composition close to the corner a_4 of triangle $a_4e_4b_4$), Δm_β will be negative and the three-phase reaction for such an alloy will be $l + \beta \rightleftharpoons \alpha$. On the other hand, alloys with smaller amounts of α will have a positive value of Δm_β and the reaction for such alloys will be $l \rightleftharpoons \alpha + \beta$. There will be a boundary line across triangle $a_4e_4b_4$ separating those alloys undergoing a peritectic reaction at this particular temperature from alloys which have the expected eutectic behaviour. The boundary line can be determined by measuring ΔX_A^α, ΔX_A^β and ΔX_A^l. In Fig. 151 $\Delta X_A^\alpha = -1$, $\Delta X_A^\beta = 3.5$ and $\Delta X_A^l = 3$ units.

$$\Delta m_\beta (X_A^\alpha - X_A^\beta) = m_\alpha \Delta X_A^\alpha + m_\beta \Delta X_A^\beta + m_l \Delta X_A^l$$
$$= -m_\alpha + 3.5\, m_\beta + 3\, m_l$$

Since

$$m_\alpha = 100 - (m_\beta + m_l)$$

then

$$\Delta m_\beta (X_A^\alpha - X_A^\beta) = -100 + 4.5\, m_\beta + 4\, m_l.$$

The boundary line corresponds to a change from the reaction $l + \beta \rightleftharpoons \alpha$ to $l \rightleftharpoons \alpha + \beta$. Along the boundary line β takes no part in the reaction, *i.e.* alloys with compositions along this line undergo a two-phase reaction $l \rightleftharpoons \alpha$ at this particular temperature. Therefore $\Delta m_\beta = 0$ along the boundary line and $-100 + 4.5\, m_\beta + 4\, m_l = 0$ is the equation of the boundary line. Using this equation, if $m_\beta = 0$ then $m_l = 20$ and if $m_l = 0$ then $m_\beta = 22.2$. The condition for $m_\beta = 0$ is that the alloy composition must lie on tie line a_4e_4 of triangle $a_4e_4b_4$. Since $m_l = 20$ the lever rule can be applied to determine the composition of an alloy with 20% liquid, given that it is composed of an α phase and a liquid phase of known co-ordinates. The composition is X_B 20.3, X_C 37.8. Similarly, the composition of the alloy on the tie line a_4b_4 for which $m_l = 0$ is X_B 27.2, X_C 34.8. The tie line joining these two points cuts off a small triangular region at the a_4 corner (Fig. 152b). In the area a_4xy the three-phase reaction at the temperature of the tie triangle $a_4e_4b_4$ is peritectic: $l + \beta \rightleftharpoons \alpha$. In the area xyb_4e_4 the reaction is eutectic: $l \rightleftharpoons \alpha + \beta$. Along the line xy the reaction is two-phase: $l \rightleftharpoons \alpha$.

Had triangle $a_5e_5b_5$ been infinitely close to $a_4e_4b_4$ the area of the peritectic region would have been confined to point a_4. In fact, a transition from a eutectic reaction to a peritectic ($l + \beta \rightleftharpoons \alpha$) begins at corner a_4—the α corner—when the tangent to the monovariant α curve, a_1a_7, at the α corner coincides with the direction of the tie line $l\alpha$ (e_4a_4). Consideration of three-phase triangles at lower temperatures will indicate that the peritectic region sweeps round from the α corner towards the β and liquid corners. At $a_5e_5b_5$ the tangent to the monovariant liquid curve, e_1e_7, at the liquid corner e_5 coincides with the tie line $l\alpha$, e_5a_5. Corresponding to this, the boundary between the peritectic and eutectic regions passes point a_5. When the temperature falls still further a stage will be reached when the tangent to the monovariant β curve, b_1b_7, at the β corner coincides with the $l\alpha$ tie line. In Fig. 151 this occurs at a temperature between isotherms $a_5e_5b_5$ and $a_6e_6b_6$. At this temperature a second peritectic region appears at the β corner of the three-phase triangle. At $a_6e_6b_6$ this second peritectic reaction area has moved into the triangle; the reactions undergone by alloys within triangle $a_6e_6b_6$ are (Fig. 152c):

(1) peritectic $l+\beta \rightleftharpoons \alpha$ in area a_6e_6ut,
(2) two-phase $l \rightleftharpoons \alpha$ along line tu,
(3) eutectic $l \rightleftharpoons \alpha+\beta$ in area $tuwv$,
(4) two-phase $l \rightleftharpoons \beta$ along line vw,
(5) peritectic $l+\alpha \rightleftharpoons \beta$ in area b_6vw.

The three-phase region ends at the degenerate tie triangle $e_7a_7(b_7)$ with the equilibrium $l_{e_7} \rightleftharpoons \alpha(\beta)_{a_7(b_7)}$. To summarise, the three-phase reaction is initially eutectic for all alloys until the temperature of the three-phase triangle $a_4e_4b_4$ is reached. From that temperature until the end of the three-phase reaction at the tie line $e_7a_7(b_7)$ the reaction type is dependent on the alloy composition within the sequence of three-phase triangles. It will either remain eutectic or become peritectic: $l+\beta \rightleftharpoons \alpha$ or $l+\alpha \rightleftharpoons \beta$. It is therefore possible to select alloys which will undergo an initial eutectic reaction followed by a peritectic reaction as they cool from the molten state. Alloy X (Fig. 151) undergoes the reactions $l \rightleftharpoons \alpha+\beta$, $l \rightleftharpoons \alpha$ and $l+\beta \rightleftharpoons \alpha$. The second reaction occurs when the boundary line between eutectic and peritectic regions meets the point X.

Isothermal sections

In the space diagram (Fig. 142) it is assumed that temperatures fall in the order $T_B > T_A > e > e_1 > T_C$.

An isothermal section at a temperature T_1, where $T_B > T_1 > T_A$, is given in Fig. 153a. It is similar to Fig. 113.

At temperatures below T_A a two-phase, $l+\alpha$, field appears in addition to the $l+\beta$ phase field shown in Fig. 153a. With further fall in temperature the two two-phase fields approach each other until, at T_2 (the binary eutectic temperature in the AB system), they meet (Fig. 153b). The liquidus isotherms l_1^1e and l_2^1e meet at the binary eutectic point e; the solidus isotherms s_1^1a and s_2^1b are joined to the points on the invariant eutectic horizontal representing the limits of solid solubility of β in α and of α in β respectively—points a and b. At a temperature below the binary eutectic an isothermal section will intersect the monovariant curves ac, bc and ee_1 (Fig. 142). At T_3 this intersection produces a three-phase triangle $a_1b_1e_1^1$, where points a_1, b_1 and e_1^1 lie on the monovariant curves ac, bc and ee_1 respectively. It will also be noticed that a two-phase region, $\alpha+\beta$, is introduced when a comparison is made of Figs. 153b and c. The $\alpha+\beta$ region is produced by the isothermal section cutting across the solubility surfaces acc_1a_1a and bcc_1b_1b to give curves a_1a_2 and b_1b_2 (a section between ab and c in Fig. 146d).

With further fall in temperature the shape of the three-phase region changes, as shown in Fig. 151, and the $\alpha+\beta$ region grows inwards. At a temperature just above, and one at the critical temperature the sections appear as illustrated in Fig. 153d and e. At the critical temperature the three-phase triangle degenerates into a straight line e_1c. The α and β phases lose their individuality and become phase $\alpha(\beta)$. The solubility isotherms a_3c and b_3c in the section taken at the critical temperature meet at point c. They form a continuous curve; similarly the solidus and liquidus isotherms form continuous curves s_1cs_2 and $l_1e_1l_2$ (cf. Figs. 153e and 144). In this way the $l+\alpha$ and $l+\beta$ phase regions are joined together as a single $l+\alpha(\beta)$ phase region.

At a temperature below the critical temperature the $l+\alpha(\beta)$ phase region is separated from

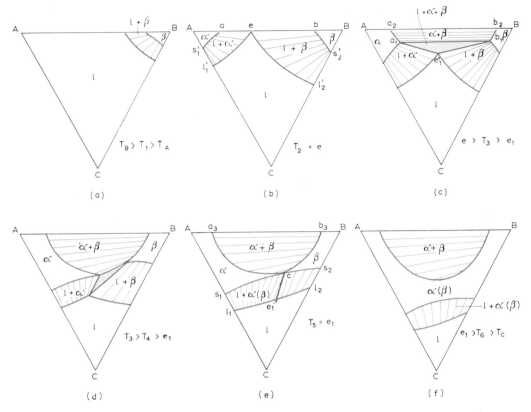

Fig. 153. Isothermal section through Fig. 142.

the $\alpha + \beta$ region (Fig. 153f) and the $l + \alpha(\beta)$ region moves closer to corner C of the concentration triangle as the temperature approaches that of component C. At T_C the $l + \alpha(\beta)$ region disappears and the section has the appearance of the room-temperature section (Fig. 142) except that the $\alpha + \beta$ region expands on cooling from T_C to room-temperature.

Vertical sections

The positions of the vertical sections are shown in Fig. 154. Since this is the first time that a vertical section passing through a three-phase region has been considered, the first section

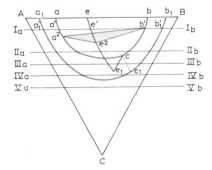

Fig. 154. Location of the vertical sections through Fig. 142.

will be treated in detail. Section Ia–Ib (Fig. 155a) is parallel and close to the side AB. When drawing vertical sections it will be found easiest at first to consider the polythermal projection (Fig. 154) together with the space diagram (Fig. 142). When experience has been gained, reference to the space diagram can be omitted.

Generally, it is best to construct vertical sections by starting with the liquidus. If a three-phase region is intersected by the section the liquidus will show a break at the point where the section intersects the monovariant liquid curve. The outline of the three-phase region can then be drawn, followed by the solidus and solubility curves. This sequence for constructing vertical sections is often the most useful to the inexperienced, and it will be followed in considering section Ia–Ib. One of the most important points to be borne in mind with vertical sections is

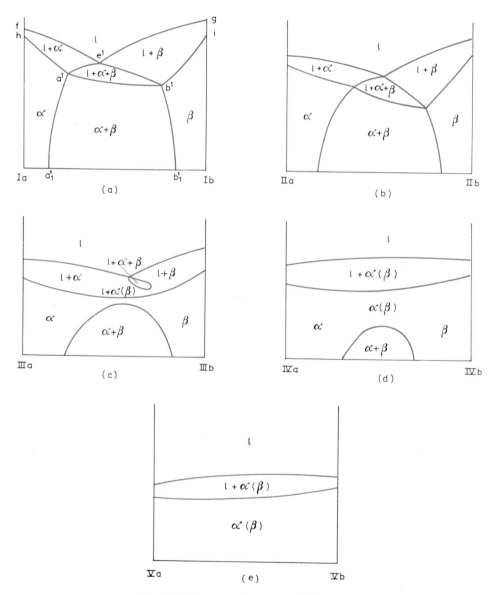

Fig. 155. Vertical sections through Fig. 142.

the correct positioning of the points where various curves meet. They should be drawn correctly with reference to both their temperature and composition.

Section Ia–Ib (Fig. 155a) contains a liquidus fe^1g. That it is of similar shape to the binary liquidus AB follows from a comparison of the space diagram (Fig. 142) with Fig. 154. Point f corresponds to the intersection of the vertical section with the liquidus of the binary AC system; point e^1 to the intersection with the monovariant liquid curve ee_1 and point g to intersection with the binary BC liquidus. Since $T_B > T_A > T_C$ point g will lie at a higher temperature than point f. Point e^1 lies at a temperature below either since $T_B > T_A > e > e_1 > T_C$. The curve fe^1 is less steep than curve e^1g. This can be confirmed by comparing the isothermals in Fig. 145a. Point e^1 is displaced slightly to the right of the eutectic composition in the binary system.

When a vertical section cuts through a three-phase region it produces a triangular region in the section. In the present case the vertical section intersects the $l\alpha$, $l\beta$ and $\alpha\beta$ ruled surfaces (*cf.* Fig. 134) to form a triangular region with curved sides. Figure 156 will assist in explaining

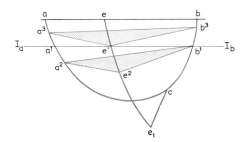

Fig. 156. Intersection of the three-phase region by section Ia–Ib.

the formation of the three-phase region in the vertical section. Considering the intersection with the $l\alpha$ surface, it should be recalled that the monovariant curve ac falls from a to c. Point a^3 therefore lies above point a^1. Point e^1 lies at the same temperature as a^3 since the triangle $a^3e^1b^3$ is an isothermal triangle. Therefore e^1 lies above a^1. In the same way we can show that e^1 lies above b^1 by consideration of the intersection with the $l\beta$ surface. Thus point e^1 lies above both points a^1 and b^1 in the three-phase region on the vertical section. The relative positions of a^1 and b^1 are best dealt with by referring to the intersection of the section with the $\alpha\beta$ ruled surface. The isothermal triangle $a^3e^1b^3$ lies above the isothermal triangle $a^2e^2b^1$. Points a^2 and b^1 lie at the same temperature. Since point a^1 lies above a^2 point a^1 must lie above point b^1. The three-phase region is therefore bounded by curves e^1a^1, e^1b^1 and a^1b^1, as shown in Fig. 155a.

Turning to the solidus it is evident that the solidus descends more steeply from the BC binary side than the AC binary side (*cf.* isotherms, Fig. 145b). The solidus descends from the binaries to the monovariant curves ac and bc (*cf.* Figs. 145b and 146a–b). Point h on the α solidus and point i on the β solidus correspond to the intersection of the vertical section with the binary AC and BC solidus curves respectively. It should be noted that the liquidus and solidus do not meet at the temperature ordinates in section Ia–Ib. Points f and h, as well as g and i, are separated on the ordinates.

Having constructed the solidus curve ha^1 and ib^1, we are left with the solubility curves. We have already seen that curve ac is the common curve for the solidus surface of α solid solution and the solubility surface of β in α solid solution. Point a^1 is on curve ac and is therefore the common point of two curves which delimit the α region. The first curve is the α solidus line ha^1 and the second the solubility curve $a^1a_1^1$ of β in α solid solution. Point a_1^1 lies on the solubility curve $a_1c_1b_1$ (Fig. 154) and the vertical section must intersect the solubility surface acc_1a_1a (Fig. 146d) along the line $a^1a_1^1$. Similarly, the vertical section intersects the solubility surface bcc_1b_1b along the line $b^1b_1^1$.

It is apparent from Fig. 155a that the vertical section through the ternary is superficially similar to the binary eutectic phase diagram AB. The two points of difference are that the liquidus and solidus do not meet on the ordinates and the binary eutectic horizontal is converted into a three-phase area in the ternary section.

Section IIa–IIb (Fig. 155b) is similar to section Ia–Ib. Certain differences are apparent. The α liquidus and solidus curves are less steep and meet the ordinate at lower temperatures. The temperature difference between liquidus and solidus is greater in section IIa–IIb (cf. the binary AC system). The same remarks apply to the β liquidus and solidus curves. The intersection of the section with the three-phase region gives a similar triangular region to that illustrated in Fig. 155a. Point e^1 is moved to the centre of the section and the difference in temperature between points a^1 and b^1 of Fig. 155a is increased in Fig. 155b. This is entirely due to the way the $\alpha\beta$ tie lines swing round as they near point c.

Section IIIa–IIIb does not pass through the $\alpha+\beta$ solubility gap acb but does cut through the solubility surfaces acc_1a_1a and bcc_1b_1b at lower temperatures than that of the degenerate tie triangle e_1c. In this way the section will show a temperature region in which all alloys are completely homogeneous. Section IIIa–IIIb is illustrated in Fig. 155c. The section cuts the monovariant liquid curve ee_1 and the degenerate tie triangle e_1c to produce the loop-shaped $l+\alpha+\beta$ region (cf. Fig. 147b and c).

Section IVa–IVb lies beyond the monovariant liquid curve ee_1 but intersects the solubility surface acc_1a_1a of β in α solid solution. It will be noted from Fig. 155d that the liquidus is now represented by a continuous smooth curve. Section Va–Vb has a similar form to Fig. 119a insofar as all alloys on the section solidify as homogeneous $\alpha(\beta)$ solid solution and undergo no solid state reactions (Fig. 155e).

Transformations in alloys on cooling

Since the vertical section Ia–Ib has been considered in detail, the transformation of alloys on cooling from the melt will be treated relative to alloy Y on section Ia–Ib (Fig. 157). On cooling alloy Y, precipitation of the α phase begins as soon as the liquid becomes saturated

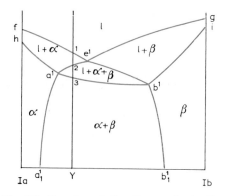

Fig. 157. Freezing of alloy Y on section Ia–Ib.

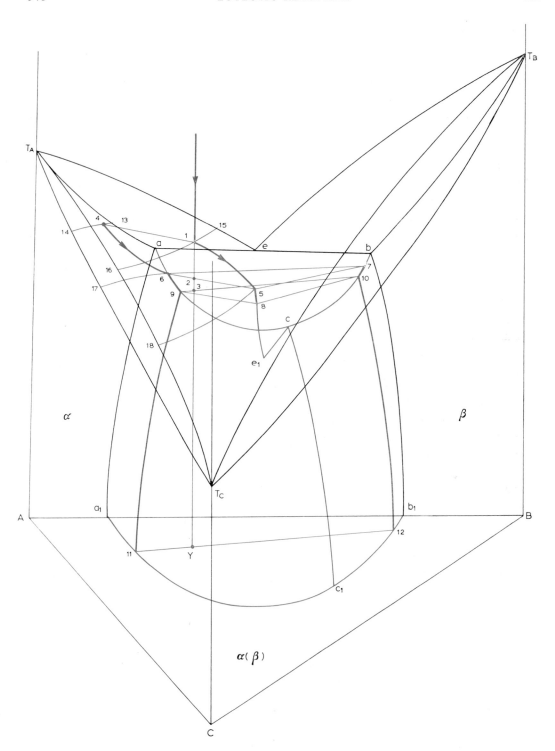

Fig. 158. Freezing of alloy *Y* illustrated with respect to the ternary space model.

with respect to α. This occurs at point 1 on the liquidus and α continues to precipitate until the liquid remaining becomes saturated with the β phase. The point at which the liquid becomes doubly saturated—with respect to both the α and β phases—is represented by point 2 on the $l\alpha$ ruled surface of the $l+\alpha+\beta$ region. From point 2 to point 3 both α and β separate from the liquid, until at point 3 the last drop of liquid is consumed. Alloy Y now consists of α and β only, and the compositions and relative amounts of each vary with further cooling to room-temperature.

Although the vertical section gives a clear idea of the sequence of crystallisation it does not give any indication of the changes in composition of the phases on cooling. For this purpose reference must be made to the space model (Fig. 142) or to its projection on the concentration triangle (Fig. 143). Until some facility has been obtained in interpreting three-dimensionally a two-dimensional projection of the space model on the concentration triangle, it is useful to consider alloy transformations with the aid of the space model. The sequence of crystallisation of alloy Y is illustrated in Fig. 158. On cooling alloy Y from the melt, solidification begins when the alloy reaches the liquidus surface at point 1. The initial composition of the α precipitating from liquid of composition 1 is represented by point 4 on the α solidus surface. The line 1–4 is the initial tie line connecting the liquid and α phases. With further cooling the liquid composition moves over the α liquidus surface along curve 1–5 and the composition of the continuously precipitated α phase moves over the α solidus surface along curve 4–6. This stage in the crystallisation of alloy Y corresponds to the interval 1–2 on the vertical section (Fig. 157). When the liquid composition reaches the monovariant liquid curve ee_1 at point 5 it becomes saturated with respect to the β solid solution and a three-phase equilibrium is set up between liquid, α and β of compositions 5, 6 and 7 respectively. With fall in temperature, both α and β are precipitated from the liquid, the α phase composition moving along curve ac from point 6 to point 9 and the β phase composition moving along curve bc from point 7 to 10 whilst the liquid composition moves along the curve ee_1 from point 5 to 8. When the liquid composition reaches 8 the last drop of liquid is consumed, as can be seen from the fact that the alloy composition, point 3, is now situated on the tie line 9–10. In accordance with the discussion on p. 155, for an alloy of composition Y, the three-phase equilibrium is eutectic. The simultaneous precipitation of α and β from the liquid over the range 5 to 8 corresponds with the interval 2–3 on the vertical section (Fig. 157).

It will be noted that the liquid composition only travels along a part of the curve ee_1, from 5 to 8, since at point 8 a temperature is reached corresponding to the completion of solidifica-

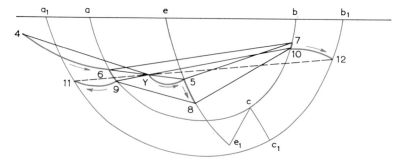

Fig. 159. Course of freezing of alloy Y illustrated by a polythermal projection of the space model (Fig. 158) on the concentration triangle.

tion of alloy Y. With the completion of solidification a two-phase equilibrium is established. The α phase of composition 9 is initially in equilibrium with the β phase of composition 10. Further fall in temperature causes an adjustment in the relative amounts of the α and β phases and in their compositions, as the α solid solution traces a curve 9–11 over the solid solubility surface acc_1a_1a of β in α solid solution, whilst the β solid solution traces a corresponding curve 10–12 over the solid solubility surface bcc_1b_1b of α in β solid solution. The surface 9–10–12–11–9 will be recognised as a skew surface. At room-temperature alloy Y is composed of α phase of composition 11 and β phase of composition 12. This latter part of the solidification sequence, corresponding to the solid state changes only, is equivalent to the interval 3–Y on the vertical section (Fig. 157).

Projection of the solidification sequence for alloy Y on the concentration triangle gives the result shown in Fig. 159. In interpreting the projection it should be remembered that point Y is identical with points 1, 2 and 3 in Fig. 158.

9.3.1.2. *A maximum critical point*

This system is similar in many respects to the system discussed at length above. The space model is illustrated in Fig. 160 and its projection on the concentration triangle in Fig. 161.

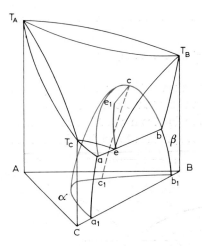

$$T_A > T_B > e_1 > T_C > e$$

Fig. 160. Space model of a ternary system with a eutectic solubility gap in one binary system and a maximum critical point on the monovariant liquid curve.

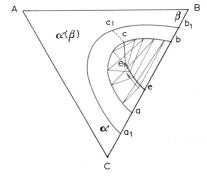

Fig. 161. Projection of the space model (Fig. 160) on the concentration triangle.

The solubility loop acb is a continuous curve rising from a and b to the critical point c. The monovariant liquid curve ee_1 rises from the binary eutectic point to e_1. As before, the line e_1c is a degenerate tie triangle. The three-phase, $l+\alpha+\beta$, region originates at this tie line and ends at the binary eutectic horizontal aeb. The three-phase equilibrium is predominantly eutectic in character except at temperatures near to the tie line e_1c. At such temperatures the equilibrium is dependent on alloy composition within the three-phase triangle, as can be proved by application of Hillert's criterion.

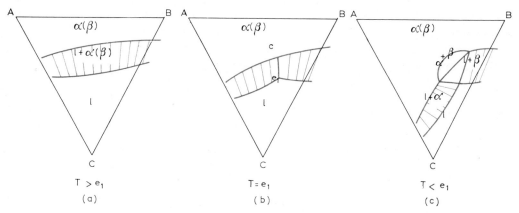

Fig. 162. Isothermal sections at temperatures (a) above e_1, (b) at e_1, and (c) below e_1.

If isothermal sections are taken at temperatures above, at and below the temperature of the degenerate tie triangle e_1c the sequence of isotherms (Fig. 162a–c) yields an example of the formation of a three-phase region from a single two-phase region.

An example of this type of system is to be found in the Au–Co–Pd system*.

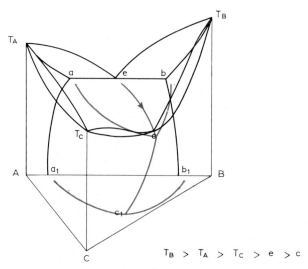

$$T_B > T_A > T_C > e > c$$

Fig. 163. Limiting case for ternary equilibria with one binary system eutectiferous.

---

* A. T. Grigoriev, E. M. Sokolovskaya, L. D. Budennaya, I. A. Iyutina and M. V. Maksimova, *Zh. Neorg. Khim.*, **1** (1956) 1052.

9.3.1.3. *A limiting case*

In the two preceding systems the solubility gap has been assumed to close within the ternary system. A limiting case arises when the solubility gap just reaches one of the binary systems (Fig. 163). This hypothetical case presupposes the existence of a minimum in the liquidus of the binary BC system. This limiting case can be looked upon as an intermediate one between that shown in Fig. 142 and the system formed when two of the binaries are eutectiferous (Fig. 164a).

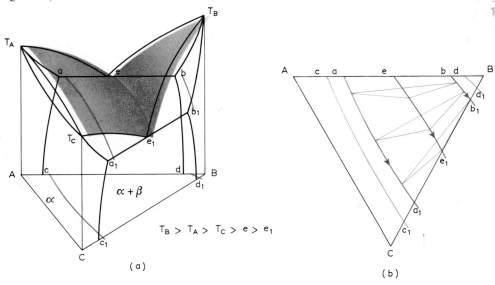

Fig. 164. Ternary system with two binary eutectic solubility gaps. (a) Space model; (b) projection on the concentration triangle.

9.3.2. *A eutectic solubility gap in two of the binary systems*

It is assumed that the third binary shows continuous series of solid solutions. The normal ternary diagram will possess a three-phase region which passes continuously from one binary eutectic horizontal to the other.

The space model of the normal system is given in Fig. 164a and the projection in Fig. 164b. The invariant points a, e and b on the AB binary system become the monovariant curves aa_1, ee_1 and bb_1 in the ternary system. The curves slope in the direction indicated by the arrows in Fig. 164b. The liquidus surface is shaded in Fig. 164a; it is divided into two by the eutectic trough ee_1

— $T_Bee_1T_B$, the liquidus surface of β solid solution,
— $T_Aee_1T_CT_A$, the liquidus surface of α solid solution.

The inter-relation of the six phase regions is illustrated in the exploded model (Fig. 165). As before, the three-phase region, $l+\alpha+\beta$, is bounded by three ruled surfaces—underneath by abb_1a_1a, on top by aee_1a_1a and ebb_1e_1e. Of course the three-phase region occupies a space in the ternary space model; curves aa_1, bb_1 and ee_1 do not lie on a plane.

The monovariant ternary eutectic reaction is very similar to the invariant binary eutectic reaction. Both involve deposition of two solid phases simultaneously from a liquid. Many writers reserve the term "ternary eutectic reaction" for a four-phase reaction of the type $l \rightleftharpoons \alpha + \beta + \gamma$ and refer to the monovariant eutectic as a "binary eutectic-type" reaction.

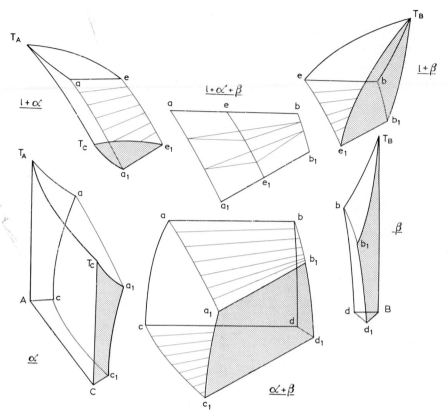

Fig. 165. The phase regions in Fig. 164a.

All alloys with compositions within the three-phase region $aebb_1e_1a_1a$ undergo a eutectic-type separation of $\alpha+\beta$ from the liquid on cooling. Alloys with compositions on the curve ee_1 solidify with typical eutectic structures; alloys in the region aee_1a_1a consist of primary dendrites of α with inter-dendritic eutectic; alloys within the region ebb_1e_1e consist of primary β with inter-dendritic eutectic.

9.4. THREE-PHASE EQUILIBRIUM INVOLVING PERITECTIC REACTIONS

By analogy with the previous eutectic systems (section 9.3) we can divide the ternary diagrams into those produced when only one binary is peritectic and those resulting from two binary peritectic diagrams. In each case the remaining binary diagrams are assumed to be of the continuous solid solution type.

9.4.1. *A peritectic solubility gap in one binary system*

A minimum or a maximum may appear in the monovariant liquid curve. This time we will consider the system with a maximum. The space model is shown in Fig. 166a and its projection on the concentration triangle in Fig. 166b. Points a, b and p represent the invariant points of the binary three-phase equilibrium $l+\beta \rightleftharpoons \alpha$ in the BC system. In the ternary system the mono-variant curve for liquid rises from point p to p_1 forming a trough in the liquidus surface,

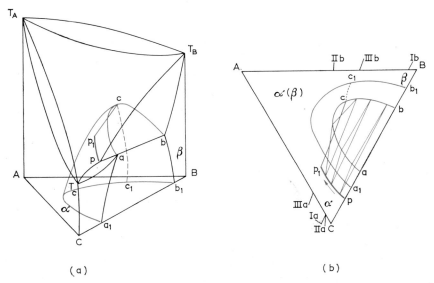

(a)　　　　　　　　　　　(b)

Fig. 166. Ternary system with a peritectic solubility gap in one binary system and a maximum critical point on the monovariant liquid curve. (a) Space model; (b) projection on the concentration triangle.

the trough petering out at p_1. The monovariant curves for α and β solid solution form a continuous curve with a critical point at c. Points p_1 and c lie at the same temperature and the line p_1c is a degenerate tie triangle.

The three-phase region is again bounded by three ruled surfaces. The upper surface is formed from a series of tie lines connecting liquid of composition along curve pp_1 with β solid solution of composition represented by points on the curve bc. There are two lower surfaces: the $l\alpha$ surface formed by tie lines between points on curves pp_1 and ac, and the $\alpha\beta$ surface formed by tie lines between points on curves ac and bc. Three isothermal sections are sufficient to define the system (Fig. 167a–c). Their construction is self-evident. The vertical section (Fig. 168a) is taken close to and parallel to the BC binary edge. The section is similar to the binary peritectic diagram. The section in Fig. 168b cuts the three-phase region obliquely to produce the loop-

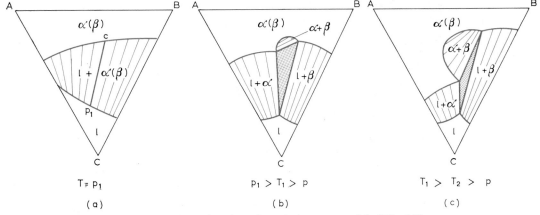

Fig. 167. Isothermal sections through the space model of Fig. 166a.

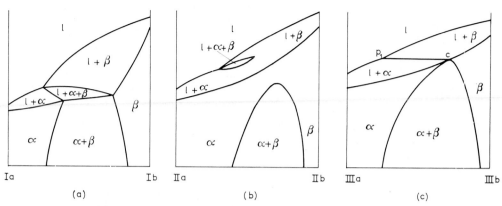

Fig. 168. Vertical sections through the space model of Fig. 166a.

shaped region of existence of $l+\alpha+\beta$. The last vertical section is assumed to pass through points p_1 and c (Fig. 168c).

Application of Hillert's criterion to the three-phase region shows that the reaction is not necessarily peritectic at temperatures just below that of the degenerate tie triangle p_1c. At lower temperatures, however, the monovariant reaction is $l+\beta \rightleftharpoons \alpha$, irrespective of alloy composition within the three-phase triangle.

If point p_1 is a minimum point on the monovariant liquid curve, the three-phase reaction $l+\beta \rightleftharpoons \alpha$ originates at the binary peritectic isotherm abp and continues as a peritectic reaction until a temperature close to the degenerate tie triangle p_1c. At such temperatures again the type of reaction is dependent on alloy composition and may be of the form $l+\beta \rightleftharpoons \alpha$, $l+\alpha \rightleftharpoons \beta$ or $l \rightleftharpoons \alpha+\beta$.

9.4.2. A peritectic solubility gap in two binary systems

This ternary system is analogous to the eutectic system described in section 9.3.2. The peritectic three-phase region may descend smoothly from one binary peritectic horizontal to the

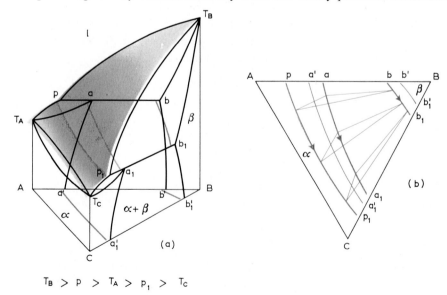

$$T_B > p > T_A > p_1 > T_c$$

Fig. 169. Ternary system with a peritectic solubility gap in two of the binary systems. (a) Space model; (b) projection on the concentration triangle.

other (Fig. 169a–b) or it may pass through a maximum or minimum. The Au–Pt–Ag system
is an example of a smooth transition from the Au–Pt peritectic to the Ag–Pt peritectic horizontal*.

9.4.3. *A transition from a binary eutectic to a binary peritectic reaction within the ternary system*

An interesting case arises when one binary is eutectic, a second peritectic, and the third isomorphous (a continuous series of solid solutions). The space diagram for this system is illustrated in Fig. 170a and the projection in Fig. 170b. As drawn, a transition occurs in the ternary
system from a monovariant peritectic reaction at temperatures close to the binary peritectic
horizontal *pab* to a monovariant eutectic reaction at temperatures close to the binary eutectic
horizontal a_1eb_1. The peritectic point p is joined to the eutectic point e; points a and a_1, representing the limits of solubility of the α solid solution, are joined together as also are points
b and b_1 for the β solid solution.

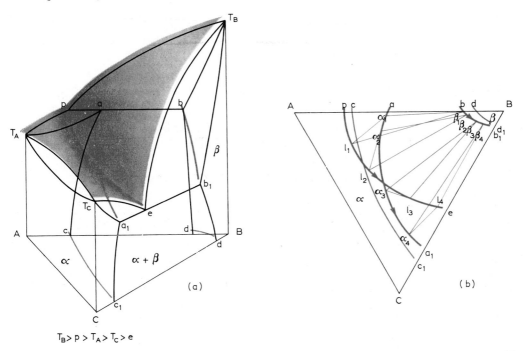

$T_B > p > T_A > T_C > e$

Fig. 170. Ternary system with a transition from a binary eutectic to a binary peritectic. (a) Space model;
(b) projection on the concentration triangle.

The monovariant curves pe and aa_1 cross over each other but curve pe always lies above
curve aa_1 in space. This is clarified by reference to Fig. 171, which shows a perspective view
of the three-phase region. The upper surface is the $l\beta$ ruled surface $pabb_1ep$; the lower surfaces
are the $l\alpha$ skew surface paa_1ep and the $\alpha\beta$ ruled surface abb_1a_1a. Tie lines are drawn on the
$l\beta$ and $l\alpha$ surfaces only.

This system was used by Hillert** to show that the transition from a peritectic to a eutectic

   * O. A. NOVIKOVA AND A. A. RUDNITSKIY, *Zh. Neorg. Khim.*, **3** (1958) 729.
  ** M. HILLERT, *J. Iron Steel Inst.*, **189** (1958) 224; *ibid.*, **195** (1960) 201.

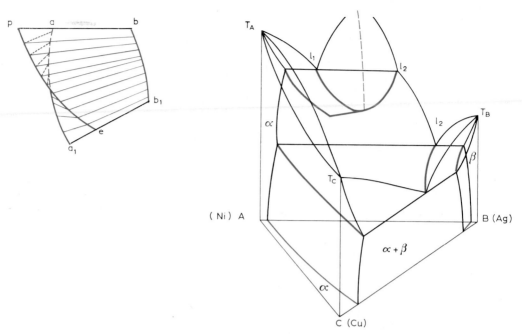

Fig. 171. The three-phase region of Fig. 170a.   Fig. 172. The Ag–Cu–Ni system (schematic).

reaction does not occur at a unique temperature (on the former basis, Fig. 149, this would be the temperature of tie triangle $\alpha_3\beta_3 l_3$, where the tangent to the monovariant liquid curve coincides with the tie line $\alpha_3 l_3$) but at different temperatures for different alloy compositions.

Ternary three-phase equilibrium has been considered with reference to binary eutectic and peritectic reactions. A whole series of ternary diagrams can be evolved by ringing the changes on binary monotectic, syntectic and metatectic reactions in combination with each other as well as with binary eutectic and peritectic reactions. As an example, Fig. 172 is a schematic illustration of the Ag–Cu–Ni* system involving a monotectic in the Ag–Ni system, a eutectic in the Ag–Cu system, and a continuous series of solid solutions in the Cu–Ni system. The ternary involves two separate three-phase equilibria,

   —that between α, l_1 and l_2, and

   —that between α, β and l_2.

---

* W. Guertler and A. Bergmann, *Z. Metallk.*, **25** (1933) 53.

CHAPTER 10

Ternary Phase Diagrams.
Four-Phase Equilibrium

A maximum of four phases may co-exist in a ternary system. The four-phase equilibria involve the transformation of one phase or an interaction between two or three phases (p. 119), for example:

$$l \rightleftharpoons \alpha + \beta + \gamma \qquad \text{ternary eutectic,}$$
$$\overline{l + \alpha} \rightleftharpoons \beta + \gamma \qquad \text{ternary quasi-peritectic,}$$
$$\overline{l + \alpha + \beta} \rightleftharpoons \gamma \qquad \text{ternary peritectic.}$$

When four phases co-exist, the equilibrium is invariant. The temperature of the transformation and the compositions of the phases taking part in it are fixed. As a result a four-phase equilibrium corresponds to the combination of four invariant points, representing the compositions of the four phases, lying on a horizontal (isothermal) plane. This plane is called the four-phase plane.

If the vapour phase is ignored there are 22 possible invariant equilibria (p. 119). In general, discussion will be limited to the three quoted above. The first represents the true ternary eutectic equilibrium as opposed to the monovariant ternary eutectic equilibrium appearing when only three phases co-exist. The second and third examples are normally considered peritectic-type equilibria. The reaction $l + \alpha + \beta \rightleftharpoons \gamma$ is a true ternary peritectic. On the other hand, the reaction $l + \alpha \rightleftharpoons \beta + \gamma$ is a mixture of eutectic and peritectic insofar as the liquid reacts with previously separated α (*i.e.* peritectic-type reaction) to produce an increase in the amount of β present plus a new phase γ (*i.e.* eutectic-type reaction). To resolve the difficulties of nomenclature it is proposed to use the term ternary quasi-peritectic for the reaction $l + \alpha \rightleftharpoons \beta + \gamma$ and ternary peritectic reaction for the reaction $l + \alpha + \beta \rightleftharpoons \gamma$.

10.1. THE EUTECTIC EQUILIBRIUM $l \rightleftharpoons \alpha + \beta + \gamma$

A ternary eutectic equilibrium involves the simultaneous separation of three solid phases from a liquid. In the most general case the liquid initially becomes saturated with respect to one phase on cooling—for example, primary crystallisation of α, corresponding to the equilibrium $l \rightleftharpoons \alpha$. At a lower temperature the liquid becomes saturated with respect to a second phase and secondary crystallisation occurs—for example, the monovariant separation of α and β according to the equilibrium $l \rightleftharpoons \alpha + \beta$. Finally, at the ternary eutectic temperature the liquid is saturated with respect to all three solid phases and solidification is completed: $l \rightleftharpoons \alpha + \beta + \gamma$.

The four-phase eutectic equilibrium must be preceded by one of the following monovariant three-phase equilibria:

$$l \rightleftharpoons \alpha + \beta$$
$$l \rightleftharpoons \alpha + \gamma$$
$$l \rightleftharpoons \beta + \gamma$$

on the assumption that the monovariant equilibria are eutectic. The simplest ternary eutectic diagram will therefore be formed from three binary eutectic systems. Such a ternary is represented by the space model and its projection (Fig. 173a–b). The four-phase plane *abc* is shaded. The ternary eutectic point *E* is a point within triangle *abc*. Points *a*, *b*, *c* and *E* correspond to the compositions of the α, β, γ and liquid phases at the eutectic temperature.

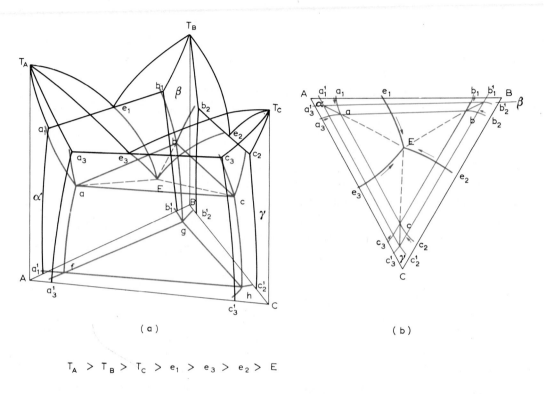

(a) (b)

$$T_A > T_B > T_C > e_1 > e_3 > e_2 > E$$

Fig. 173. Ternary eutectic system. (a) Space model; (b) projection on the concentration triangle.

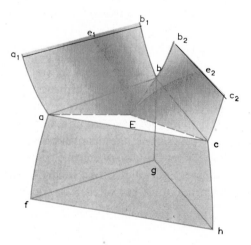

Fig. 174. The eutectic four-phase plane as the junction of four tie triangles.

The space model looks rather complex at first glance. It can best be understood by recalling that the four-phase plane is the junction of four tie triangles. From above, these are the $l\alpha\beta$, $l\alpha\gamma$ and $l\beta\gamma$ tie triangles; from below, the $\alpha\beta\gamma$ tie triangle. The monovariant three-phase equilibrium $l \rightleftharpoons \alpha + \beta$ originates at the AB binary eutectic horizontal $a_1 e_1 b_1$ and ends at the four-phase plane in the tie triangle aEb (Fig. 174). Similarly, the $l \rightleftharpoons \beta + \gamma$ equilibrium begins at

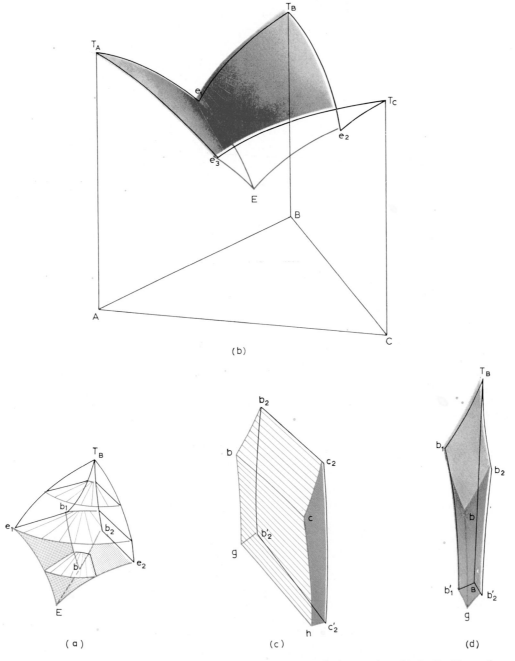

Fig. 175. Phase regions in the space model of Fig. 173a. (a) The $l+\beta$ phase region; (b) the liquidus surface; (c) the $\beta+\gamma$ phase region; (d) the β phase region.

the BC eutectic $b_2e_2c_2$ and ends at the tie triangle bEc on the four-phase plane. The $l \rightleftharpoons \alpha + \gamma$ equilibrium runs from the AC binary eutectic $a_3e_3c_3$ to aEc on the four-phase plane. In Fig. 174 this particular three-phase equilibrium is omitted to allow a clear view of the $l\alpha\beta$ and $l\beta\gamma$ phase regions. The $\alpha\beta\gamma$ region extends from the four-phase plane abc to the concentration triangle fgh.

The three regions of primary crystallisation, $l \rightleftharpoons \alpha$, $l \rightleftharpoons \beta$, and $l \rightleftharpoons \gamma$ sit on top of the monovariant three-phase regions. The region of primary crystallisation of β is shown in Fig. 175a; it stretches over the trough between the $l\alpha\beta$ and $l\beta\gamma$ three-phase regions. The shape of the

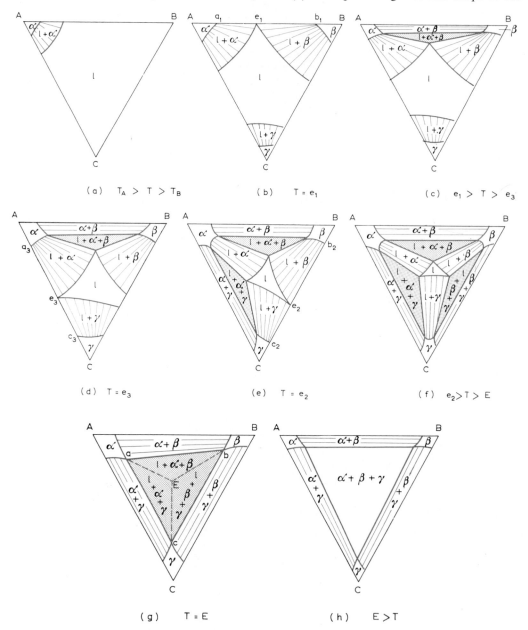

Fig. 176. Isothermal sections through the space model of Fig. 173a.

$l+\beta$ phase region is clarified by the three isothermal sections shown in Fig. 175a. Each region of primary crystallisation touches the other two along the appropriate monovariant eutectic curves, the $l+\beta$ region contacting the $l+\alpha$ region along the curve e_1E and the $l+\gamma$ region along curve e_2E. The liquidus surface (Fig. 175b) is therefore formed of three surfaces separated by the eutectic troughs e_1E, e_2E and e_3E.

In addition to the three two-phase regions involving the liquid phase there also exist the two-phase regions $\alpha+\beta$, $\beta+\gamma$ and $\alpha+\gamma$. These are situated between the corresponding three-phase regions and the binary faces of the space model; the $\beta+\gamma$ region (Fig. 175c) touches the $l+\beta+\gamma$ phase region along the $\beta\gamma$ ruled surface $b_2c_2cbb_2$ and the $\alpha+\beta+\gamma$ phase region along the $\beta\gamma$ ruled surface $bchgb$. The vertical face $b_2^1b_2c_2c_2^1b_2^1$ forms the $\beta+\gamma$ miscibility gap in the BC binary system.

Only the single phase regions remain. The β phase region (Fig. 175d) is a solid solution of A and C in B; it is therefore located at the B corner. It makes contact with the two-phase regions $\alpha+\beta$, $\beta+\gamma$ and $l+\beta$ over surfaces $b_1bgb_1^1b_1$, $b_2b_2^1gbb_2$ and $T_Bb_1bb_2T_B$ respectively; with the three-phase regions $l+\alpha+\beta$, $l+\beta+\gamma$ and $\alpha+\beta+\gamma$ along the curves b_1b, b_2b and bg respectively.

Isothermal sections

Sections taken through the space model (Fig. 173a) are given in Fig. 176a–h. The development of the three-phase region $l+\alpha+\beta$ (Fig. 176a–c), is comparable with Fig. 153a–c. At the AC binary eutectic temperature (Fig. 176d) the liquidus isotherms of the α and γ phases meet at the eutectic point e_3. When a section is taken at the temperature of the BC eutectic (Fig. 176e) the liquid phase region assumes a triangular shape, one apex of which is the eutectic point e_2. At a temperature just above the ternary eutectic temperature there are three three-phase regions (Fig. 176f). At the ternary eutectic temperature the section appears as in Fig. 176g. In the present figure the troughs between the three-phase regions have been denoted by the dotted lines Ea, Eb and Ec. At the ternary eutectic temperature the liquid phase region has shrunk from a triangle (Fig. 176f) to the eutectic point E. Below this temperature solidification is complete and the sections have the appearance shown in Fig. 176h.

Vertical sections

All sections (Figs. 177–178) are taken parallel to the AB edge. Section 1–2 (Fig. 178a) will be considered. Figure 179 indicates how this section is constructed relative to the space model. Point w lies on curve e_1E and marks the separation of the regions of primary crystallisation

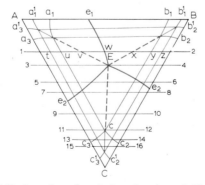

Fig. 177. Location of vertical section through Fig. 173a.

of α and β—the $l+\alpha$ and $l+\beta$ phase regions. Curves wv and wx are traces of the intersection of the section with the $l\alpha$ and $l\beta$ ruled surfaces of the $l+\alpha+\beta$ phase region; points v and x lie on the troughs aE and bE separating neighbouring three-phase regions. From v the section intersects the $l\alpha$ ruled surface of the $l+\alpha+\gamma$ phase region to give the curve vr, point r being on the eutectic horizontal $a_3e_3c_3$. The lower $\alpha\gamma$ ruled surface of the $l+\alpha+\gamma$ phase region is cut along the curve ru. Points u, v and x all lie on the four-phase plane. On the other side of the section the intersection of the $l+\beta+\gamma$ phase region by the section gives the curves xs and sy. Point y is also on the four-phase plane, and the line $uvxy$ is horizontal as a consequence. From points u and y the vertical section intersects the $\alpha+\beta+\gamma$ phase region along the $\alpha\gamma$ and $\beta\gamma$ ruled surfaces respectively to produce curves ut and yz.

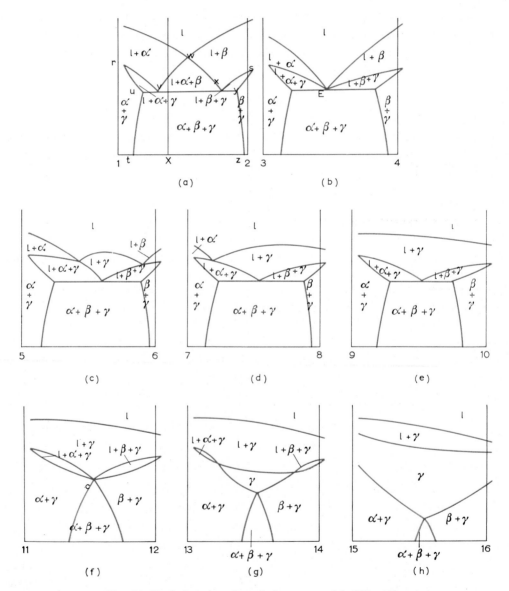

Fig. 178. Vertical sections through the space model of Fig. 173a.

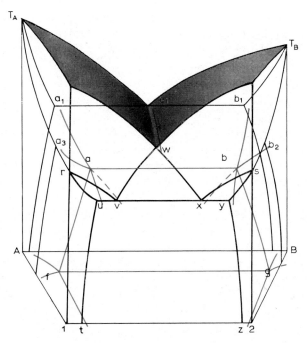

Fig. 179. Construction of vertical section 1–2.

Transformations in alloys on cooling

There are four different patterns of behaviour for alloys which undergo the ternary eutectic reaction, *i.e.* alloys whose composition lies within the triangle *abc* (Fig. 173a–b).

(1) The alloy of composition E solidifies isothermally to produce a completely ternary eutectic structure (Fig. 180).

(2) Alloys whose compositions lie on those parts of the monovariant liquid curves e_1E, e_2E and e_3E within triangle *abc*, initially precipitate two solid phases ($\alpha + \beta$, $\beta + \gamma$ or $\alpha + \gamma$) until the temperature falls to the ternary eutectic temperature. The remaining liquid then solidifies as the ternary eutectic.

(3) Alloys which lie on the troughs (tie lines) aE, bE and cE will initially undergo primary crystallisation of α, β or γ until the liquid composition has moved over the liquidus surface directly to E. The ternary eutectic reaction follows directly.

(4) Alloys within triangles abE, bcE and acE undergo primary, secondary and tertiary crystallisation. In the area abE the sequence of crystallisation would involve the primary separation of α or β (depending on whether the alloy composition was to the left or right of curve e_1E), secondary crystallisation of $\alpha + \beta$, followed by tertiary crystallisation of $\alpha + \beta + \gamma$ at the eutectic temperature.

The freezing of the fourth type of alloy is illustrated in Fig. 181. Solidification of alloy X begins with the precipitation of α of composition α_1 and follows the normal patterns until the liquid meets the monovariant curve e_1E. At this temperature liquid of composition l_2 is in equilibrium with α of composition α_2. Since α_2 is on the curve a_1a, this α_2 is in equilibrium with β of composition β_2, *i.e.* a three-phase equilibrium between l_2, α_2 and β_2 is established. The secondary precipitation of $\alpha + \beta$ from the liquid continues until the liquid composition reaches the eutectic point, E. The tie triangle $\alpha_2\beta_2l_2$ descends from l_2 to the tie triangle *abc* on the four-phase plane.

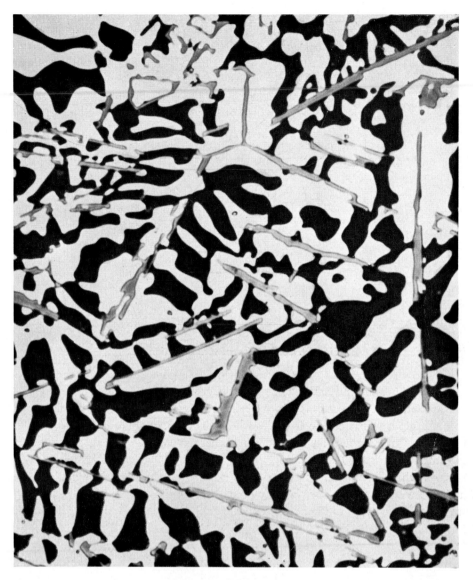

Fig. 180. Microstructure of the ternary eutectic in the Al–Cu–Si system. α light, θ dark, Si grey, $\times 900$ (courtesy Metallurgical Services).

The remaining liquid now decomposes eutectically to yield $\alpha + \beta + \gamma$ of composition a, b and c respectively. From the four-phase plane to room temperature only changes in the relative amounts of the solid α, β and γ phases occur.

Alloy X is assumed to lie on the vertical section 1–2. Reference to Fig. 178a indicates that this alloy passes through the phase regions $l/l+\alpha/l+\alpha+\beta/\alpha+\beta+\gamma$ on cooling, in agreement with the sequence deduced from Fig. 181.

The Ga–In–Sn system* is an example of a ternary eutectic system.

---

* H. SPENGLER, *Z. Metallk.*, **46** (1955) 464.

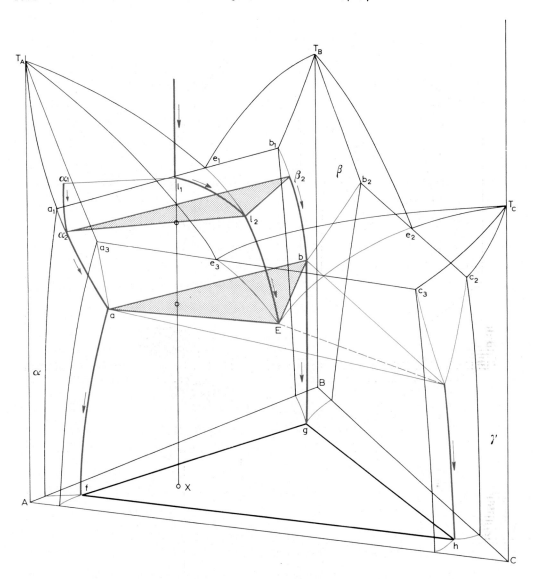

Fig. 181. Freezing of alloy X illustrated with reference to the ternary space model.

10.2. VARIANTS OF THE TERNARY EUTECTIC DIAGRAM

In discussing the preceding system we assumed that the three binary systems were of the eutectic type. This is not a necessary condition for the formation of a ternary eutectic. The binary systems may involve one or more peritectic types. The three-phase monovariant region in the ternary system would start at the peritectic horizontal and end at the appropriate tie triangle on the four-phase plane. A three-dimensional model is given in Fig. 182 to illustrate the point for one binary peritectic system (*cf.* Fig. 171). A ternary eutectic can be produced with one, two or three binary peritectic systems.

A four-phase equilibrium could occur in a ternary composed of two binary eutectics, with the third binary showing a complete series of solid solutions. This can happen if the eutectic valley

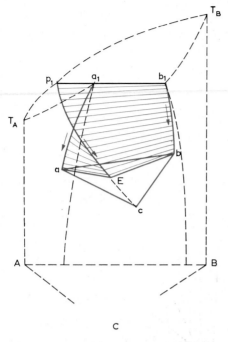

Fig. 182. Variant of the ternary eutectic system in which one binary is a peritectic.

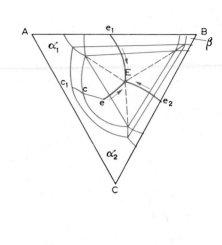

Fig. 183. Ternary eutectic system in which two of the binaries are eutectics.

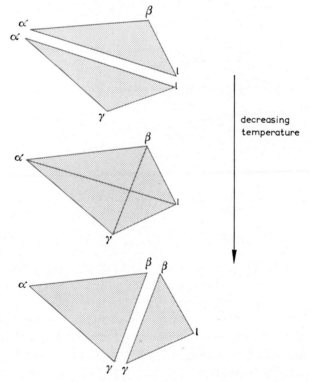

decreasing temperature

Fig. 184. Equilibrium relationships in a quasi-binary peritectic system on cooling through the invariant reaction at the four-phase plane.

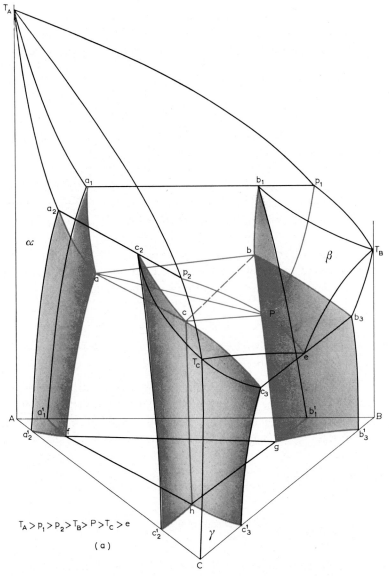

$$T_A > P_1 > P_2 > T_B > P > T_C > e$$

(a)

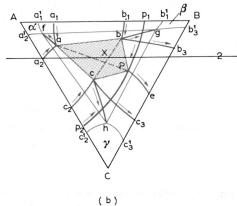

(b)

Fig. 185. Ternary quasi-peritectic system.
(a) Space model; (b) projection on the
concentration triangle.

from e to E (Fig. 183) peters out on the liquidus surface at e^*. The behaviour of the monovariant liquid curve eE is identical to the case previously illustrated in Figs. 160–161, excepting that the three-phase region ends on the four-phase plane instead of the BC binary eutectic horizontal.

10.3. THE QUASI-PERITECTIC EQUILIBRIUM $l+\alpha \rightleftharpoons \beta+\gamma$.

We have already noted on p. 125 that the ternary quasi-peritectic reaction $l+\alpha \rightleftharpoons \beta+\gamma$ involves the appearance of two three-phase monovariant equilibria at temperatures above the invariant four-phase plane and two different three-phase monovariant reactions below this plane. A schematic representation of the sequence of events on cooling through the four-phase plane is given in Fig. 184. The three-phase equilibria $l\alpha\beta$ and $l\alpha\gamma$ fall to the four-phase plane; reaction then occurs such that the three-phase equilibria leaving the four-phase plane are $l\beta\gamma$ and $\alpha\beta\gamma$.

The ternary systems in which a quasi-peritectic reaction occurs can be differentiated on the basis of the type of equilibrium found in the two monovariant reactions which precede the invariant reaction. Both monovariant reactions—$l\alpha\beta$ and $l\alpha\gamma$—can be peritectic, one can be eutectic or both can be eutectic.

10.3.1. *Both three-phase monovariant equilibria preceding the quasi-peritectic reaction are peritectic*

If the $l\alpha\beta$ and $l\alpha\gamma$ equilibria are peritectic and the $l\beta\gamma$ equilibrium is eutectic, the space model would have the form shown in Fig. 185a. The $l+\alpha \rightleftharpoons \beta$ peritectic of the AB system is joined to the tie triangle abP on the four-phase plane $abPc$ and the $l+\alpha \rightleftharpoons \gamma$ peritectic of the AC system to the tie triangle acP of the four-phase plane. The four-phase plane is the junction of four tie triangles:

abP	peritectic $l\alpha\beta$ equilibrium ⎫	descending to the four-phase plane;
acP	peritectic $l\alpha\gamma$ equilibrium ⎭	
bcP	eutectic $l\beta\gamma$ equilibrium ⎫	descending from the four-phase plane.
abc	$\alpha\beta\gamma$ equilibrium ⎭	

The diagonal aP represents all possible mixtures of liquid and α phase which can form all possible mixtures of β and γ phase, represented by the other diagonal bc. From the four-phase plane the $l\beta\gamma$ equilibrium, represented by the tie triangle Pbc, descends to the eutectic horizontal c_3eb_3 of the BC system. The $\alpha\beta\gamma$ equilibrium descends from tie triangle abc to the room temperature tie triangle fgh. The various phase regions are shown separately in Fig. 186; the ruled surfaces of the monovariant three-phase regions are individually identified.

The α liquidus surface is Ap_1Pp_2A, the β liquidus Bp_1PeB and the γ liquidus Cp_2PeC (Fig. 185b). Only alloys whose compositions lie on the four-phase plane $abPc$ undergo the quasi-peritectic reaction. Taking these alloys, we can distinguish two cases: alloys within triangle abP and those within triangle acP. The first group initially separate primary α on cooling from the liquid state until the liquid composition meets the curve p_1P. The $l\alpha\beta$ equilibrium is then

---

* Further details are given by W. HUME-ROTHERY, J. W. CHRISTIAN AND W. B. PEARSON, *Metallurgical Equilibrium Diagrams*, The Institute of Physics, London, 1952, p. 259.

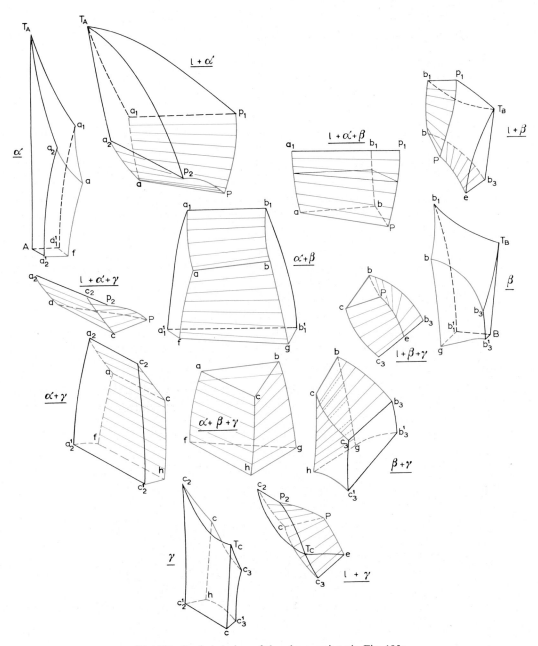

Fig. 186. Exploded view of the phase regions in Fig. 185a.

established. Basically, this is a monovariant peritectic reaction $l+\alpha \rightleftharpoons \beta$. The liquid reacts with some of the primary α to form the new phase β. This reaction proceeds until the liquid reaches the ternary quasi-peritectic point P. At this temperature the four-phase reaction occurs. In the area abX all the liquid is consumed in the reaction $l+\alpha \rightleftharpoons \beta+\gamma$ and we are left with only $\alpha+\beta+\gamma$ of composition a, b and c respectively. In the area bPX of triangle abP all the α is consumed by reaction with the liquid and we are left with $l+\beta+\gamma$ after the quasi-peritectic reaction. The liquid then deposits $\beta+\gamma$ eutectically as it moves down the curve Pe.

The reactions undergone by alloys within the triangle acP are similar to the type of reactions discussed for alloys in triangle abP. Summarising:

| Triangle abP | area abX | (1) Primary separation of α, $l \rightleftharpoons \alpha$. |

(2) Secondary crystallisation of $\alpha + \beta$ by reaction of liquid with some of the primary α, $l + \alpha \rightleftharpoons \beta$.

(3) Invariant reaction, $\underline{l} + \alpha \rightleftharpoons \beta + \gamma$.

(4) All liquid consumed, leaving $\alpha + \beta + \gamma$ equilibrium stable to room temperature.

area bXP (1) (2) as in area abX.

(3) Invariant reaction, $l + \alpha \rightleftharpoons \beta + \gamma$.

(4) All α (the primary α) consumed, leaving $l + \beta + \gamma$ equilibrium.

(5) Solidification completed by monovariant eutectic reaction $l \rightleftharpoons \beta + \gamma$.

Triangle acP area acX (1) Primary crystallisation of α, $l \rightleftharpoons \alpha$.

(2) Secondary crystallisation of $\alpha + \gamma$ by reaction of liquid with some of the primary α, $l + \alpha \rightleftharpoons \gamma$.

(3) Invariant reaction, $\underline{l} + \alpha \rightleftharpoons \beta + \gamma$.

(4) All liquid consumed, leaving $\alpha + \beta + \gamma$ equilibrium stable to room temperature.

area cXP (1) (2) as in area acX.

(3) Invariant reaction, $l + \underline{\alpha} \rightleftharpoons \beta + \gamma$.

(4) All α consumed, leaving $l + \beta + \gamma$ equilibrium.

(5) Solidification completed by monovariant eutectic reaction, $l \rightleftharpoons \beta + \gamma$.

Tie line aP section aX (1) Primary crystallisation of α, $l \rightleftharpoons \alpha$.

(2) No monovariant reaction since the liquid composition falls directly to P and these alloys then undergo the invariant reaction $\underline{l} + \alpha \rightleftharpoons \beta + \gamma$.

(3) All liquid is consumed in producing composition X (a mixture of $\beta + \gamma$), leaving the $\alpha + \beta + \gamma$ equilibrium stable to room temperature.

section XP (1) Primary crystallisation of α, $l \rightleftharpoons \alpha$.

(2) Liquid again reaches P directly: $l + \underline{\alpha} \rightleftharpoons \beta + \gamma$.

(3) All α consumed, giving a $l + \beta + \gamma$ equilibrium.

(4) Solidification completed by eutectic separation of $\beta + \gamma$, $l \rightleftharpoons \beta + \gamma$.

alloy X (1) Primary crystallisation of α, $l \rightleftharpoons \alpha$.

(2) As above, but in the invariant reaction both the liquid and α are consumed, leaving a mixture of $\beta + \gamma$ of composition b and c after reaction.

(3) On cooling from the four-phase plane to room temperature, α is precipitated from the $\beta + \alpha$ to give an $\alpha + \beta + \gamma$ equilibrium.

The lever rule can be used (p. 122) to calculate the changes in the amount of the phases as they undergo the invariant reaction.

Isothermal sections

Isothermal sections taken above, at, and below the temperature of the four-phase plane are given in Fig. 187a–c.

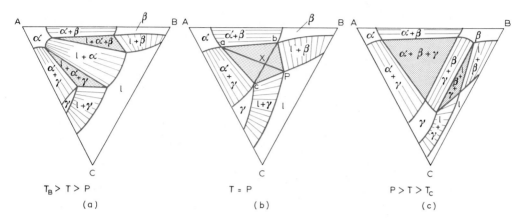

Fig. 187. Isothermal sections through the space model of Fig. 185a.

Vertical sections

A vertical section from a_2, parallel to the AB edge (Fig. 185b) is given in Fig. 188a–b. The section intersects the α and β liquidus surfaces to give the curve 1 2 3, point 2 lying on the monovariant liquid curve p_1P. It also intersects the four-phase plane $abcP$ to give the horizontal line 4 5 6 7. Points 4, 5, 6 and 7 lie on the tie lines ac, aP, bc and bP respectively. The section cuts the four three-phase regions $l+\alpha+\gamma$, $l+\alpha+\beta$, $\alpha+\beta+\gamma$ and $l+\beta+\gamma$ —the first two above the four-phase plane and the latter two below the plane. In each case the vertical section cuts across the appropriate ruled surfaces of the three-phase regions. The $l+\alpha+\beta$ region in Fig. 188 is produced by intersection with the $l\alpha$ ($a_1p_1Paa_1$) and $l\beta$ ($b_1p_1Pbb_1$) ruled surfaces of the $l+\alpha+\beta$ region (Figs. 185a and 186). This intersection is illustrated in Fig. 188c. The reader is left to exercise his three-dimensional perception by verifying the correctness of the other three-phase regions in Fig. 188a–b by reference to Figs. 185 and 186. Similarly, to gain practice in thinking three-dimensionally the reader is advised to consider other vertical sections in the manner illustrated above for section a_2–2.

Tabular representation of ternary equilibria

A schematic representation of ternary equilibria, found in German literature particularly, interlinks the binary and ternary reactions in tabular form. The ternary eutectic system (Fig. 173a) can be represented as shown in Table 3.

The binary invariant reactions are located vertically according to the temperature at which they occur. The ternary reaction(s) are subsequently plotted at the appropriate temperature(s) and linked to the binary reactions as shown. As each ternary invariant reaction is associated with four three-phase equilibria a check can be made to ensure that the system has been interpreted correctly.

The ternary quasi-peritectic system given in Fig. 185a can be represented as shown in Table 4.

The ternary peritectic equilibrium to be considered later (Fig. 193) can be tabulated as shown in Table 5.

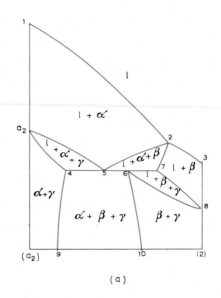

(a)

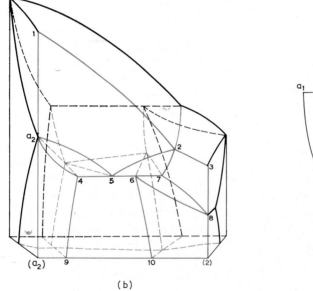

(b)

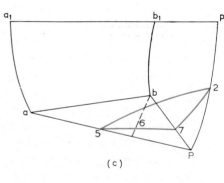

(c)

Fig. 188. A vertical section through the space model of Fig. 185a. (a) The vertical section a_2–2; (b) construction of the vertical section; (c) intersection of the vertical section with the $l+\alpha+\beta$ phase region.

TABLE 3

Binary AB	Ternary	Binary AC	Binary BC
$l \rightleftharpoons \alpha + \beta$		$l \rightleftharpoons \alpha + \gamma$	$l \rightleftharpoons \beta + \gamma$
	$\boxed{l \rightleftharpoons \alpha + \beta + \gamma}$		
	$\alpha + \beta + \gamma$		

TABLE 4

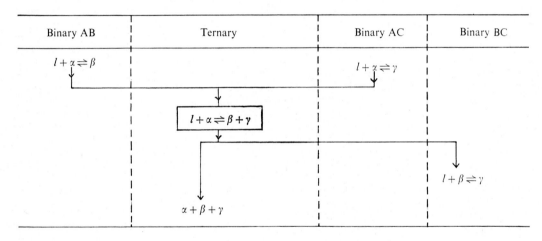

Binary AB	Ternary	Binary AC	Binary BC
$l + \alpha \rightleftharpoons \beta$		$l + \alpha \rightleftharpoons \gamma$	
	$\boxed{l + \alpha \rightleftharpoons \beta + \gamma}$		
			$l + \beta \rightleftharpoons \gamma$
	$\alpha + \beta + \gamma$		

TABLE 5

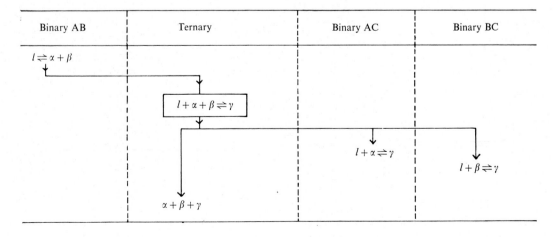

Binary AB	Ternary	Binary AC	Binary BC
$l \rightleftharpoons \alpha + \beta$			
	$\boxed{l + \alpha + \beta \rightleftharpoons \gamma}$		
		$l + \alpha \rightleftharpoons \gamma$	$l + \beta \rightleftharpoons \gamma$
	$\alpha + \beta + \gamma$		

10.3.2. *One of the three-phase monovariant equilibria preceding the quasi-peritectic reaction is eutectic and one peritectic*

Since there is little difference between this system and the preceding one, its consideration is left to the reader. It is best constructed from a eutectic AB system and a peritectic AC system. A more interesting case occurs when the AB system contains an intermediate phase formed by peritectic reaction (*cf.* Fig. 68). The space model (Fig. 189) has been drawn with some of the

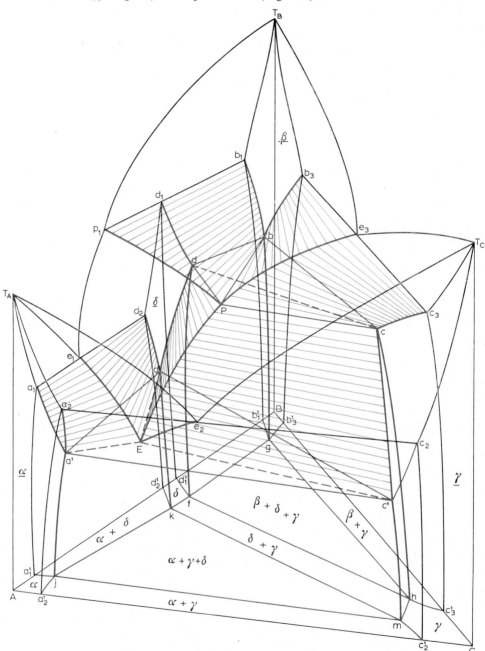

Fig. 189. Ternary system involving an incongruently-melting binary intermediate phase.

ruled surfaces of the monovariant three-phase regions individually marked. This ternary system can be regarded as a combination of a ternary quasi-peritectic diagram and a ternary eutectic diagram. The equilibria involved can tabulated as shown in Table 6.

The liquidus surface (Fig. 189) is divided into four parts: $T_B p_1 P e_3 T_B - \beta$ liquidus, $T_C e_3 P E e_2 T_C - \gamma$ liquidus, $T_A e_1 E e_2 T_A - \alpha$ liquidus and $p_1 P E e_1 p_1 - \delta$ liquidus. Of the two-phase regions the $l+\delta$ is of interest in that it stretches below the $l\delta$ ruled surface, $p_1 d_1 dP p_1$, of the $l\beta\delta$ phase

TABLE 6

Binary AB	Ternary	Binary AC	Binary BC
$l + \beta \rightleftharpoons \delta$		$l \rightleftharpoons \beta + \gamma$	
	$l + \beta \rightleftharpoons \gamma + \delta$		
	$l \rightleftharpoons \delta + \gamma$		
$l \rightleftharpoons \alpha + \delta$			$l \rightleftharpoons \alpha + \gamma$
	$l \rightleftharpoons \alpha + \gamma + \delta$		
$\beta + \gamma + \delta$	$\alpha + \gamma + \delta$		

region. An alloy whose composition fell within the area $p_1 d_1 dP p_1$ would pass through the phase regions $l \,/\, l+\beta \,/\, l+\beta+\delta \,/\, l+\delta$ on cooling. Its subsequent freezing would depend on whether it met the $l\delta$ ruled surface $e_1 d_2 d^1 E e_1$ of the $l\alpha\delta$ region, the $l\delta$ ruled surface $Pdd^1 EP$ of the $l\gamma\delta$ region, or the solidus surface $d_1 dd^1 d_2 d_1$ of δ. Irrespective of this, the alloy would separate primary β; this would then react with the liquid to form δ until all the primary β was consumed. The alloy would then consist of $l+\delta$. It now enters the $l+\delta$ phase region and separates more δ. This second separation of δ is a primary crystallisation of δ compared to the prior formation of δ as a result of the monovariant peritectic reaction, $l+\beta \rightleftharpoons \delta$.

The quasi-peritectic diagram and the ternary eutectic diagram are separated by the near-vertical $\gamma\delta$ ruled surface $dchfd$. The vertical section which intersects point d_1 on the AB binary, the tie lines db and Pc, and proceeds to the C corner is given in Fig. 190. It emphasises the fact that the section from the δ phase to component C is not a quasi-binary section.

An example of this type of system is the Au–Ge–Sb ternary* in which the δ phase is the intermediate phase $AuSb_2$.

---

* G. ZWINGMANN, Z. Metallk., **55** (1964) 192.

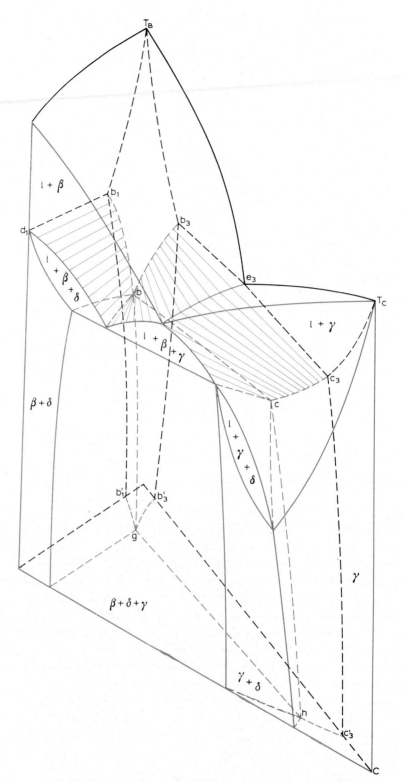

Fig. 190. A vertical section through the space model of Fig. 189.

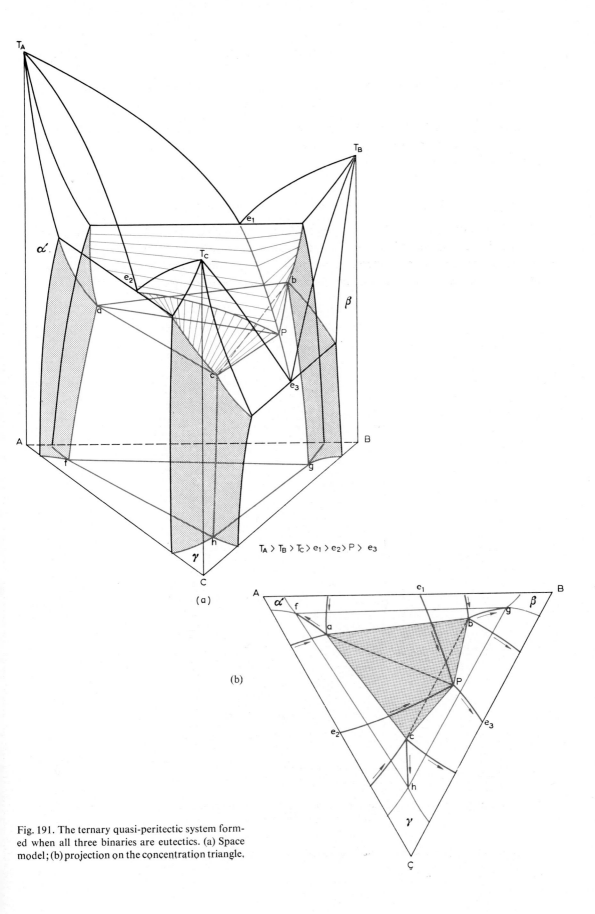

$T_A > T_B > T_C > e_1 > e_2 > P > e_3$

Fig. 191. The ternary quasi-peritectic system formed when all three binaries are eutectics. (a) Space
model; (b) projection on the concentration triangle.

10.3.3. *Both of the three-phase monovariant equilibria preceding the quasi-peritectic reaction are eutectic*

A ternary quasi-peritectic reaction is possible when all three binaries are eutectic (Fig. 191). For example the binary Re–Ta and Re–W eutectics fall to a quasi-peritectic four-phase plane in the Re–Ta–W* system.

10.4. THE TERNARY PERITECTIC EQUILIBRIUM $l+\alpha+\beta \rightleftharpoons \gamma$

In the formation of a new solid phase γ by reaction between a liquid and two other solid phases α and β, it must be assumed that the α and β phases are formed from the liquid. The ternary peritectic equilibrium must therefore be preceded by a three-phase monovariant reaction between liquid, α and β. This reaction can be either eutectic or peritectic. Once the four-phase

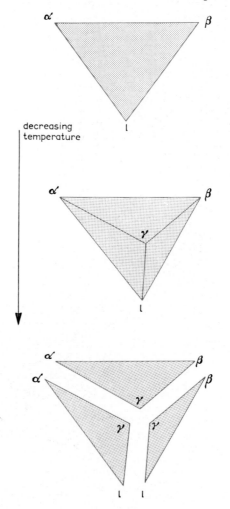

Fig. 192. Equilibrium relationships in a ternary peritectic system on cooling through the invariant reaction at the four-phase plane.

* J. H. BROPHY, M. H. KAMDAR AND J. WULFF, *Trans. AIME*, **221** (1961) 1137.

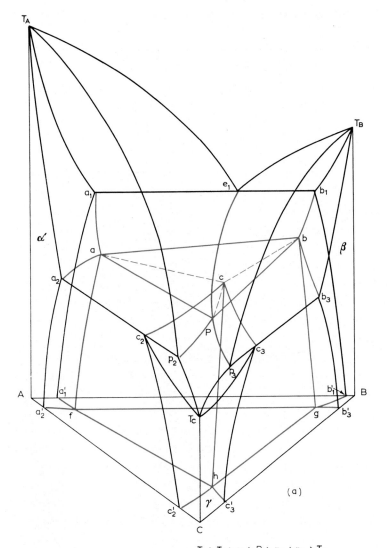

$T_A > T_B > e_1 > P > p_2 > p_3 > T_C$

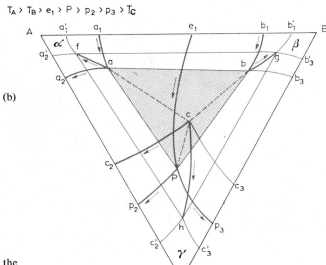

Fig. 193. Ternary peritectic system.
(a) Space model; (b) projection on the
concentration triangle.

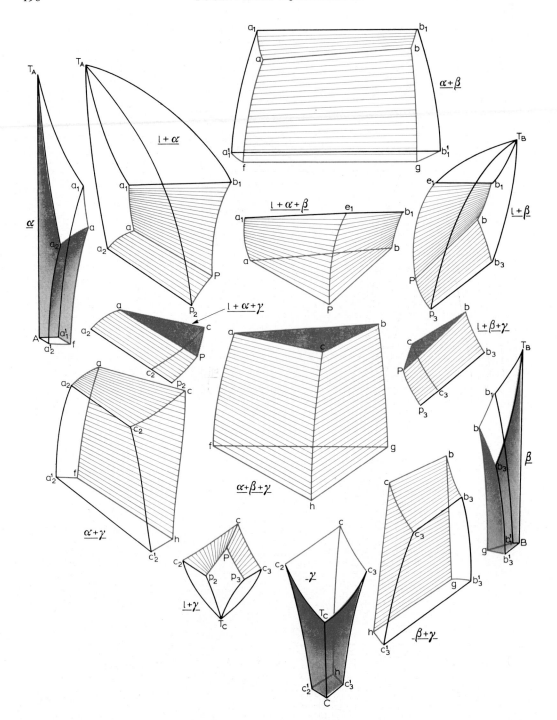

Fig. 194. Exploded view of the phase regions in Fig. 193a.

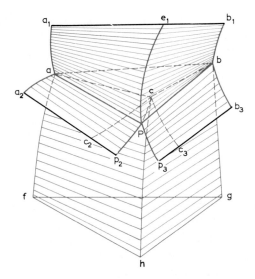

Fig. 195. The ternary peritectic four-phase plane as the junction of four tie triangles.

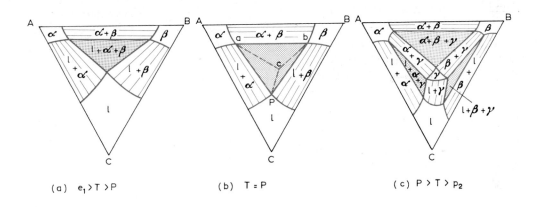

(a) $e_1 > T > P$ (b) $T = P$ (c) $P > T > P_2$

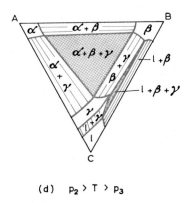

(d) $P_2 > T > P_3$

Fig. 196. Isothermal sections through the space model of Fig. 193a.

reaction is completed a three-phase equilibrium will follow. As the four-phase plane is the junction of four three-phase equilibria, three three-phase equilibria will follow the invariant reaction. A schematic representation of the sequence of events on cooling through the four-phase plane is given in Fig. 192. The $\alpha\beta\gamma$, $l\alpha\gamma$ and $l\beta\gamma$ equilibria descend from the four-phase plane to lower temperatures. As an example of this type of system, we will briefly consider the situation arising when the reaction preceding the four-phase ternary peritectic is eutectic, $l \rightleftharpoons \alpha+\beta$, and the subsequent $l\alpha\gamma$ and $l\beta\gamma$ equilibria are peritectic, $l+\alpha \rightleftharpoons \gamma$ and $l+\beta \rightleftharpoons \gamma$. The space model and its projection are given in Fig. 193 and the individual phase regions in Fig. 194. The junction of the three-phase equilibria on the four-phase plane is illustrated in Fig. 195.

A few isothermal sections are given at various temperatures through the four-phase plane in Fig. 196. They illustrate the conversion of the $l\alpha\beta$ eutectic equilibrium above P to the three equilibria $\alpha\beta\gamma$, $l\alpha\gamma$ and $l\beta\gamma$ below P. In a sense this transition is the image of the ternary eutectic equilibrium (*cf.* Fig. 176h, g and f with Fig. 196a, b and c). Reference to the space model, its projection and the exploded model of the phase regions (Fig . 193 and 194) should allow the reader to construct vertical sections and trace the course of freezing of any ternary in this system.

This type of ternary reaction occurs in the system C–Ti–W*. The partial system W–TiC–Ti behaves exactly as the system ABC in Fig. 193a. There is a quasi-binary eutectic formed between W and TiC, a peritectic in the W–Ti system and a peritectic in the TiC–Ti system. The ternary peritectic reaction is $l+W+TiC \rightleftharpoons Ti$. Another example occurs in the Al–Mg–Zn system** with the ternary peritectic reaction $l+\delta+\lambda \rightleftharpoons \gamma$ at 393 °C. In this case, the δ phase is Mg_3Al_2 and the λ and γ are ternary intermediate phases. This paper by Clark contains the first photomicrographic work on the structure of ternary peritectics.

* H. NOWOTNY, E. PARTHÉ, R. KIEFFER AND F. BENESOVSKY, *Z. Metallk.*, **45** (1954) 97.
** J. B. CLARK, *Trans. Am. Soc. Metals*, **53** (1961) 295.

CHAPTER 11

Ternary Phase Diagrams.
Intermediate Phases

Intermediate phases may melt congruently or incongruently. They may occur as either binary or ternary phases.

11.1. CONGRUENTLY-MELTING INTERMEDIATE PHASES

11.1.1. *Binary intermediate phases*

The simplest system is one in which only one binary intermediate phase occurs. Assume the AB system contains an intermediate phase δ. The ternary will contain the five phases α, β, γ, δ and liquid l. Since the maximum number of phases which can co-exist is four, there must be more than one four-phase equilibrium in the ternary. The possibilities are:

$$l \rightleftharpoons \alpha + \beta + \gamma$$
$$l \rightleftharpoons \alpha + \beta + \delta$$
$$l \rightleftharpoons \alpha + \gamma + \delta$$
$$l \rightleftharpoons \beta + \gamma + \delta.$$

The more usual combination is of the last two equilibria, implying equilibrium in the solid state between $\alpha\gamma\delta$ and $\beta\gamma\delta$. This can be envisaged if there is direct equilibrium between γ and δ, splitting the ternary system into two partial systems AδC and BδC (Fig. 197a). It often happens that the δ phase forms a quasi-binary system with component C. The ternary phase diagram

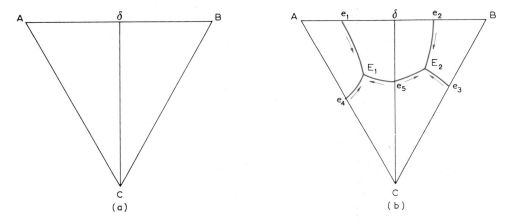

Fig. 197. The effect of a congruent intermediate phase, (a) splitting the ternary system into two partial systems AδC and BδC, (b) in forming eutectic systems associated with a saddle point on the quasi-binary section δC.

would then contain two ternary eutectic points, E_1 and E_2; the quasi-binary section divides
the ternary ABC into two simple ternary eutectic systems (Fig. 197b) in the same way that a
congruently-melting intermediate phase divides a binary diagram into two (Fig. 78). Each
partial ternary system, AδC and BδC, is identical to the ternary eutectic system given in Fig.
173. The monovariant liquid curves only are given in Fig. 197b. It will be noted that the eutectic

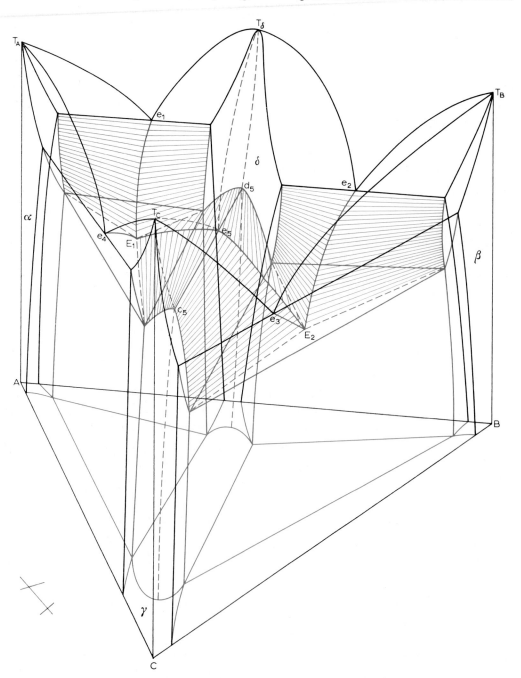

Fig. 198. Ternary space model for the system represented in Fig. 197b.

point e_5 on the quasi-binary section δC is a saddle point. The curve $E_1e_5E_2$ passes through a maximum at e_5, whereas the liquidus curve $T_\delta e_5 T_C$ passes through a minimum at e_5. This is in agreement with the statement on p. 148 that the temperature is a maximum or a minimum if three phases—δ, γ and liquid—can be represented by points on a straight line. The straight line in question is the quasi-binary eutectic horizontal $c_5e_5d_5$ (Fig. 198).

The space model for this system is given in Fig. 198. The quasi-binary section is shown dotted. To the left of the section is the ternary eutectic system between α, δ and γ; to the right the ternary eutectic system between β, δ and γ. In view of the close similarity with Fig. 173 this system need not be considered further. Examples of this type of system are to be found in the Al–Mg–Si and Al–Sb–Zn systems. In the former system a quasi-binary eutectic section is formed between Al and the intermediate phase Mg$_2$Si (Fig. 199). The Al–Mg$_2$Si eutectic occurs at 595 °C and 8.15 %Mg, 4.75 %Si. The ternary eutectic E_2 in the Al–Mg–Mg$_2$Si partial system is more complex but there is a ternary eutectic E_1 at 451 °C between Al, Mg$_2$Si and Al$_3$Mg$_2$. The eutectic E_1 contains 33.2 %Mg and 0.37 %Si.

The solid solubility isothermals of the γ phase (a solid solution of Mg and Si in Al) have the

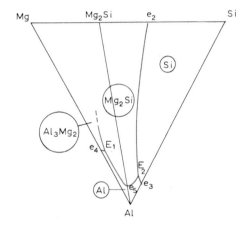

Fig. 199. The Al–Mg–Si system (schematic).

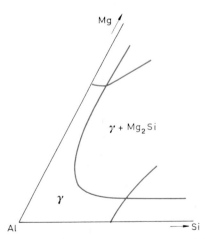

Fig. 200. Solid solubility isotherm for the Al-based solid solution in the Al–Mg–Si system (schematic).

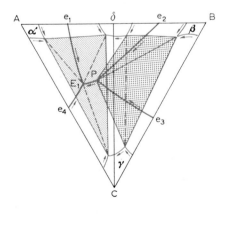

Fig. 201. A congruent intermediate phase associated with a quasi-peritectic ternary reaction.

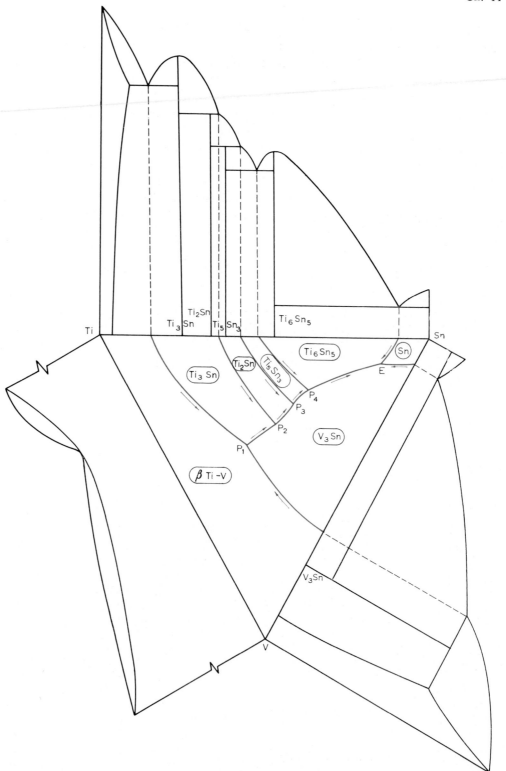

Fig. 202. The Sn–Ti–V system (schematic). (After W. Köster and K. Haug, *Z. Metallk.*, **48** (1957) 327; courtesy Dr. Riederer-Verlag GmbH.)

form shown in Fig. 200. According to Hume-Rothery*, this type of isothermal is characteristic of the case in which an intermediate phase (Mg_2Si) separates on crossing the boundary from the single-phase γ to the two-phase $\gamma + Mg_2Si$ region. The separation of Mg_2Si from the γ phase is regarded as a reaction in which the Law of Mass Action can be applied:

$$2Mg + Si \rightleftharpoons Mg_2Si$$

$$K = \frac{(a_{Mg})^2(a_{Si})}{(a_{Mg_2Si})}$$

where a_{Mg}, *etc.*, denote activities of the phases. As the activity of Mg_2Si along the solubility isothermal is constant,

$$(a_{Mg})^2(a_{Si}) = \text{constant}$$

or,

$$2 \log a_{Mg} + \log a_{Si} = \text{constant}.$$

On the assumption that $X_{Mg} \simeq a_{Mg}$ and $X_{Si} \simeq a_{Si}$, a plot of $\log(\text{at.} \%Mg)$ against $\log(\text{at.} \%Si)$ should produce a linear relation with slope -2. In fact, along the γ solid solubility isothermal there is a simple relation between the at.%Mg and Si which is reasonably close** to that predicted by the above argument.

The liquidus isothermals within the primary crystallisation region of Mg_2Si can also be analysed in this way***.

It is not essential for the section δC to be quasi-binary. The type of diagram illustrated in Fig. 201 is also possible. In this case the eutectic originating at e_2 ends at a peritectic point P which is located in the triangle AδC. The solid equilibria are again $\alpha\delta\gamma$ and $\beta\delta\gamma$. Examples of congruently-melting intermediate phases which are involved in quasi-peritectic ternary reactions are to be found in the literature. The system Sn–Ti–V† can be used as an example of this, as well as a good example of a series of quasi-peritectic reactions. The monovariant liquid curves are shown in Fig. 202. The ternary invariant reactions are:

P_1	$l + \beta$ Ti-V	$\rightleftharpoons Ti_3Sn + V_3Sn$
P_2	$l + \quad Ti_3Sn$	$\rightleftharpoons Ti_2Sn + V_3Sn$
P_3	$l + \quad Ti_2Sn$	$\rightleftharpoons Ti_5Sn_3 + V_3Sn$
P_4	$l + \quad Ti_5Sn_3$	$\rightleftharpoons Ti_6Sn_5 + V_3Sn$
E	l	$\rightleftharpoons Ti_6Sn_5 + V_3Sn + Sn.$

As mentioned initially, there is the possibility of the combination of the $l \rightleftharpoons \alpha + \beta + \gamma$ and $l \rightleftharpoons \alpha + \beta + \delta$ equilibria (Fig. 203). There is a saddle point at e_5 but no quasi-binary section in this system.

If more than one binary intermediate phase exists, the ternary system will be correspondingly more complex. Simplification can be obtained by splitting the system ABC into a series of partial systems, as was done in Fig. 197a. This division of the concentration triangle of a complicated ternary system into a series of simple triangles is called triangulation of the system‡.

---

   * W. HUME-ROTHERY, *Phil. Mag.*, **22** (1936) 1013; see also H. K. HARDY, *J. Inst. Metals*, **81** (1952) 432.

  ** W. HUME-ROTHERY AND G. V. RAYNOR, *The Structure of Metals and Alloys*, Monograph No. 1, 4th edn., Institute of Metals, London, 1962, p. 161.

 *** W. HUME-ROTHERY, *Phil. Mag.*, **22** (1936) 1013; also H. W. L. PHILLIPS, *J. Inst. Metals*, **72** (1946) 229.

   † W. KÖSTER AND K. HAUG, *Z. Metallk.*, **48** (1957) 327.

   ‡ Further details are given by A. PRINCE, *Met. Rev.*, **8** (1963) 213.

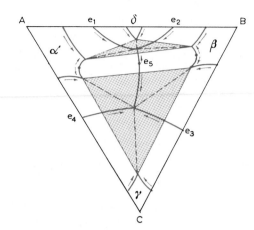

Fig. 203. A congruent intermediate phase associated with a saddle point but no quasi-binary section.

An example of this for the case where each binary system contains one intermediate phase, X, Y and Z, is given in Fig. 204a; in Fig. 204b the monovariant liquid curves are projected on the concentration triangle on the assumption that all reactions are eutectic. Binary and quasi-binary eutectic points are numbered 1, 2, 3, ..., 9; ternary eutectic points are numbered I, II, III and IV. There are four different systems possible:

(1) binary equilibrium between $X-Z$, $X-Y$ and $Y-Z$
(2) binary equilibrium between $X-Z$, $X-Y$ and $X-C$
(3) binary equilibrium between $X-Z$, $Z-B$ and $X-Z$
(4) binary equilibrium between $Y-A$, $X-Y$ and $Y-Z$

The As–Ga–Zn* system is one in which there are three congruent phases, but in this case two occur on one binary (As–Zn) and the third on the As–Ga binary. The Ga–Zn system is a eutectic one. Figure 205 is a projection of the monovariant liquid curves on the concentration triangle. There are three quasi-binary sections—Zn–GaAs, GaAs–Zn_3As_2 and GaAs–$ZnAs_2$—and four ternary eutectic points. The invariant reaction temperatures are as follows:

e_1	quasi-binary eutectic $l \rightleftharpoons GaAs + Zn$	at	414 °C,
e_2	quasi-binary eutectic $l \rightleftharpoons GaAs + Zn_3As_2$	at	972 °C,
e_3	quasi-binary eutectic $l \rightleftharpoons GaAs + ZnAs_2$	at	754 °C,
E_1	ternary eutectic $l \rightleftharpoons GaAs + Zn + Ga$	at	$\sim$ 20 °C,
E_2	ternary eutectic $l \rightleftharpoons GaAs + Zn + Zn_3As_2$	at	$\sim$410 °C,
E_3	ternary eutectic $l \rightleftharpoons GaAs + Zn_3As_2 + ZnAs_2$	at	$\sim$750 °C,
E_4	ternary eutectic $l \rightleftharpoons GaAs + ZnAs_2 + As$	at	$\sim$720 °C.

The region in which GaAs is the primary phase to crystallise from the liquid is lightly shaded. It illustrates the dominating behaviour of the high-melting phase GaAs in this system. For clarity no solid solubility between any of the phases has been indicated.

A tabular presentation of the ternary equilibria in the As–Ga–Zn system is given in Table 7.

Kurnakov** has pointed out a direct relationship between the principal elements in systems containing only binary congruent intermediate phases:

$$k = E = c_2 + 1 = q + 1 = m + 1,$$

* W. KÖSTER AND W. ULRICH, Z. Metallk., **49** (1958) 361.
** N. S. KURNAKOV, Collected Works, Vol. 1, Acad. Sci. S.S.S.R., Moscow, 1961.

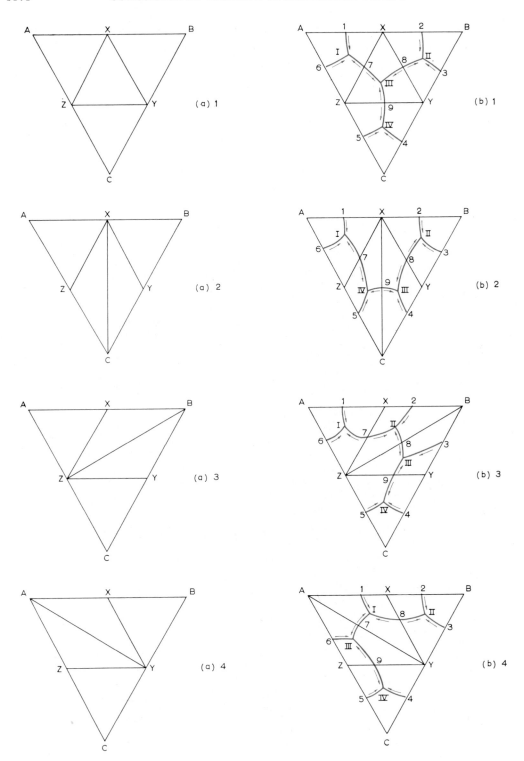

Fig. 204. Triangulation of ternary systems, (a) containing congruent intermediate phases on each binary system, (b) the corresponding equilibria for eutectic reactions.

where k, E, c_2, q and m are the number of secondary triangles, ternary eutectic points, binary congruent intermediate phases, quasi-binary sections, and saddle points respectively. When only ternary congruent intermediate phases exist the relation has the form:

$$k = E = 2c_3 + 1 = 2/3\,q + 1 = 2/3\,m + 1,$$

where c_3 is the number of ternary phases. If both binary and ternary congruent intermediate phases are present the relation is:

$$k = E = 1 + c_2 + 2c_3 = q + 1 - c_3$$
$$= m + 1 - c_3.$$

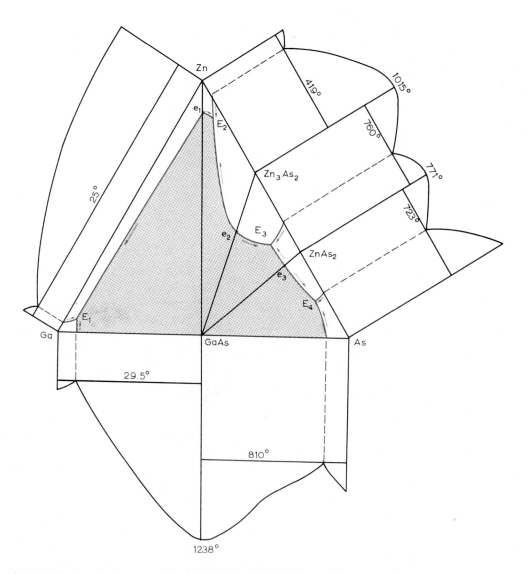

Fig. 205. The As–Ga–Zn system (schematic). (After W. KÖSTER AND W. ULRICH, *Z. Metallk.*, **49** (1958) 361; courtesy Dr. Riederer-Verlag GmbH.)

TABLE 7

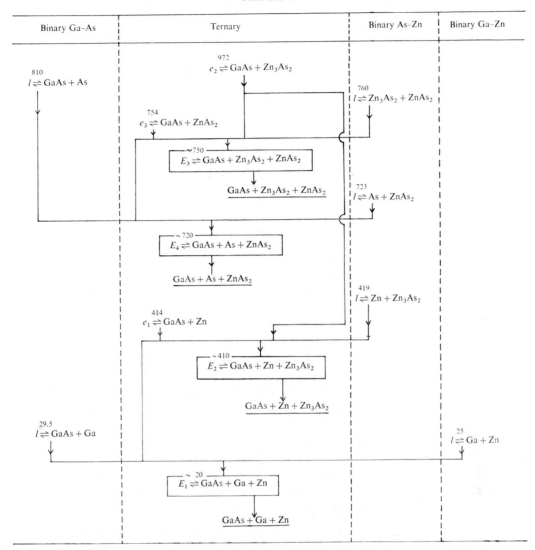

The four three-phase equilibria underlined are stable down to room-temperature.

Rhines[*] has noted that the relation $k = 1 + c_2 + 2c_3$ can be used to check ternary isothermal sections irrespective of whether they contain congruent or incongruent phases, provided we define k as the number of three-phase tie triangles, c_2 as the number of single-phase regions joined to a binary edge (but excluding the α, β and γ terminal solid solutions based on components A, B and C), and c_3 as the number of single-phase regions completely within the ternary system. Taking as an example the system in Fig. 204b1 at a temperature just above the lowest-melting of the ternary eutectics, say ternary eutectic III, we would obtain the isothermal section of Fig. 206. There are three single-phase regions joined to the binaries (δ, ε and η phase regions) and one single-phase region within the ternary (liquid l).

[*] F. N. RHINES, *Phase Diagrams in Metallurgy*, McGraw-Hill, New York, 1956, p. 201.

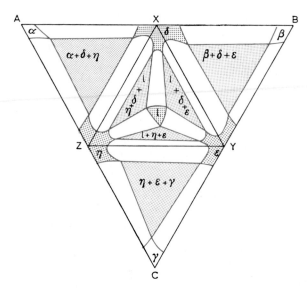

Fig. 206. Isothermal section through Fig. 204b (1) at a temperature just above the lowest melting ternary eutectic (III).

Therefore the number of tie triangles, $k = 1 + c_2 + 2c_3 = 1 + 3 + 2(1) = 6$. These are $\alpha + \delta + \eta$, $\beta + \delta + \varepsilon$, $\eta + \varepsilon + \gamma$, $l + \delta + \eta$, $l + \delta + \varepsilon$ and $l + \eta + \varepsilon$.

If any of the single-phase regions stretch across the triangle ABC from one binary to another, it is necessary to subtract one tie triangle from the result calculated by the above formula for each continuous phase region. The Kurnakov and Rhines' rules are useful in checking the construction of ternary systems and their isothermal sections when intermediate phases are involved[*].

11.1.2. *Ternary intermediate phases*

Ternary intermediate phases are fairly common in alloy systems. The appearance of a ternary congruent intermediate phase is best considered by reference to an actual system. Incongruently-melting ternary phases are dealt with later.

The first published account of a ternary alloy phase referred to the discovery of the congruent phase Hg_2KNa[**]. The more recent work of Dobbener on the Bi–Cu–Mg system will be taken as a typical example of a system with a ternary congruent phase, BiCuMg[***]. A projection of the monovariant liquid curves on the concentration triangle is given in Fig. 207. It will be noted that the phase region of primary crystallisation of the BiCuMg ternary phase is totally enclosed within the ternary. This region is lightly shaded. The temperatures of the binary invariant reactions are given on the binary systems erected on the edges of the ternary in Fig. 207. The temperatures of the ternary invariant reactions are:

e_1	quasi-binary eutectic $l \rightleftharpoons$ Bi + BiCuMg	at	~ 271 °C
e_2	quasi-binary eutectic $l \rightleftharpoons$ Cu + BiCuMg	at	690 °C

   [*] C. S. Smith (*Metal Interfaces*, Am. Soc. Metals, Cleveland, Ohio, 1952, pp. 65–108) has applied topological methods to the checking of ternary isothermal sections. No account of this approach is included since it is felt that the Kurnakov and Rhines' rules provide a more fruitful approach.
  [**] E. Jänecke, *Z. Physik. Chem.* (*Leipzig*), **57** (1906) 507; also *Z. Metallk.*, **20** (1928) 113.
[***] R. Dobbener (data of W. Rose), *Z. Metallk.*, **48** (1957) 413.

e_3	quasi-binary eutectic $l \rightleftharpoons Bi_2Mg_3 + BiCuMg$	at	695 °C
e_4	quasi-binary eutectic $l \rightleftharpoons Bi_2Mg_3 + Cu_2Mg$	at	655 °C
e_5	quasi-binary eutectic $l \rightleftharpoons Bi_2Mg_3 + CuMg_2$	at	557 °C
E_1	ternary eutectic $l \rightleftharpoons Bi + Cu + BiCuMg$	at	265 °C
E_2	ternary eutectic $l \rightleftharpoons Bi + Bi_2Mg_3 + BiCuMg$	at	255 °C
E_3	ternary eutectic $l \rightleftharpoons Cu + Bi_2Mg_3 + BiCuMg$	at	630 °C
E_4	ternary eutectic $l \rightleftharpoons Bi_2Mg_3 + Cu_2Mg + CuMg_2$	at	546 °C
E_5	ternary eutectic $l \rightleftharpoons Mg + Bi_2Mg_3 + CuMg_2$	at	470 °C
P	quasi-peritectic $l + Cu_2Mg \rightleftharpoons Cu + Bi_2Mg_3$	at	660 °C

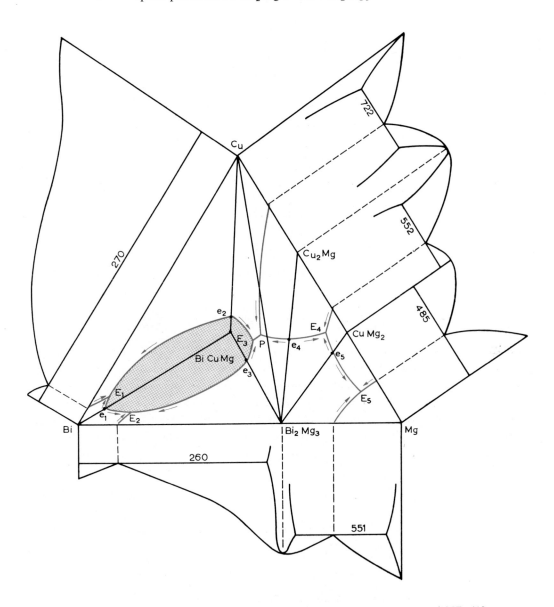

Fig. 207. The Bi–Cu–Mg system (schematic). (After R. Dobbener, *Z. Metallk.*, **48** (1957) 413; courtesy Dr. Riederer-Verlag GmbH.)

The ternary phase BiCuMg melts at 735 °C. The data given above allows the system to be constructed in the form of a three-dimensional model. Since the prime interest is in the ternary phase, Fig. 208 has been included as a representation of the shape of the solid BiCuMg phase region. At room-temperature it has a hexagonal shape: $c_1^1 o_2^1 c_2^1 o_1^1 c_3^1 o_3^1$. Curves $T_{BiCuMg}c_3 c_3^1$, $T_{BiCuMg}c_2 c_2^1$ and $T_{BiCuMg}c_1 c_1^1$ are the boundaries for the existence of the BiCuMg phase in the quasi-binary sections BiCuMg–Bi_2Mg_3, BiCuMg–Cu and BiCuMg–Bi respectively. Points o_1, o_2, and o_3 are the invariant points for the BiCuMg phase on the ternary eutectic planes $Bi_2Mg_3 + Cu + BiCuMg$ (630 °C), $Bi + Cu + BiCuMg$ (265 °C), and $Bi + Bi_2Mg_3 + BiCuMg$ (255 °C) respectively.

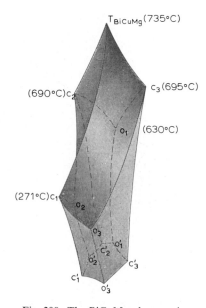

Fig. 208. The BiCuMg phase region.

A similar state of affairs is found in the Cu–Mg–Sb system*. The congruent phase CuMgSb melts at 940 °C, the difference of 200 °C when compared with the melting point of BiCuMg reflecting the greater stability of the CuMgSb phase. Indeed, the quasi-binary eutectic temperatures at e_2 and e_3 (Fig. 207) are only about 40 °C below the melting point of BiCuMg. In this sense the CuMgSb phase can be thought of as being more representative of a ternary congruent phase, and the BiCuMg phase as being a step towards a system with an incongruently-melting ternary phase.

11.2. INCONGRUENTLY-MELTING INTERMEDIATE PHASES

11.2.1. *Binary intermediate phases*

The case where an intermediate phase occurs in one of the three binary systems has already been considered on p. 190 and in Fig. 189. If an intermediate phase occurs in two of the binary

---

* R. DOBBENER AND R. VOGEL, *Z. Metallk.*, **50** (1959) 412, and an interpretation by E. JÄNECKE (*Kürzgefasstes Handbuch aller Legierungen*, 2nd edn., Otto Spamer Verlag, Leipzig, 1949) of data by E. SCHEIL AND W. SIBERT, *Z. Metallk.*, **33** (1941) 389.

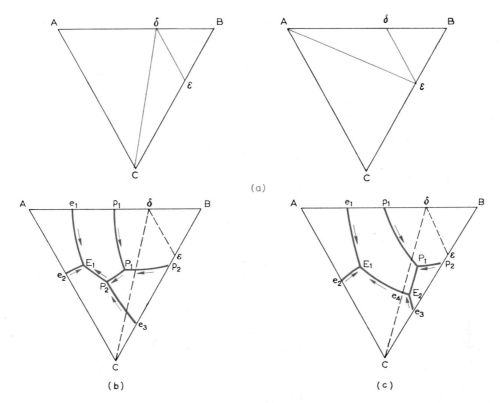

Fig. 209. Ternary system formed when two of the binaries contain incongruent intermediate phases. (a) Alternative equilibria possible; (b) equilibria when the quasi-peritectic point P is located in the partial system $A\delta C$; (c) in the partial system $C\delta\varepsilon$.

systems, the ternary system ABC can be triangulated in two ways (Fig. 209a). It will be assumed that two-phase equilibrium is established between δ and ε and between δ and C. Two cases can be distinguished (Fig. 209b and c).

The first case involves two quasi-peritectic and one eutectic equilibria:

$$
\begin{array}{ll}
P_1 & l+\beta \rightleftharpoons \delta+\varepsilon \\
P_2 & l+\varepsilon \rightleftharpoons \delta+\gamma \\
E_1 & l \rightleftharpoons \alpha+\delta+\gamma.
\end{array}
$$

The AB and BC binaries are similar to the AB system in Fig. 189; the AC binary is a simple eutectic. In tabular form the ternary can be represented as shown in Table 8.

The second case merits further discussion. The quasi-peritectic point P_1 is located in the partial system $\delta\varepsilon C$. This leads to the following invariant equilibria:

$$
\begin{array}{ll}
P_1 & l+\beta \rightleftharpoons \delta+\varepsilon \\
E_2 & l \rightleftharpoons \delta+\varepsilon+\gamma \\
E_1 & l \rightleftharpoons \alpha+\delta+\varepsilon \\
e_4 & l \rightleftharpoons \delta+\gamma.
\end{array}
$$

The section from C to δ would be a true quasi-binary section but for the overhang of the $l+\beta+\delta$ phase region. To clarify the various invariant reactions Fig. 210 is drawn with the

TABLE 8

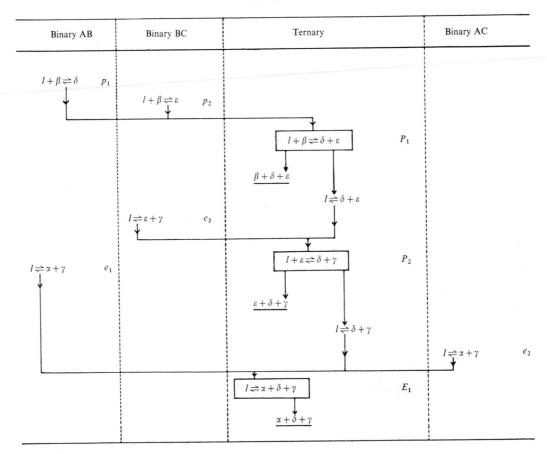

Binary AB	Binary BC	Ternary	Binary AC

$l + \beta \rightleftharpoons \delta \qquad p_1$

$l + \beta \rightleftharpoons \varepsilon \qquad p_2$

$\boxed{l + \beta \rightleftharpoons \delta + \varepsilon} \qquad P_1$

$\underline{\beta + \delta + \varepsilon}$

$l \rightleftharpoons \delta + \varepsilon$

$l \rightleftharpoons \varepsilon + \gamma \qquad e_3$

$\boxed{l + \varepsilon \rightleftharpoons \delta + \gamma} \qquad P_2$

$\underline{\varepsilon + \delta + \gamma}$

$l \rightleftharpoons \alpha + \gamma \qquad e_1$

$l \rightleftharpoons \delta + \gamma$

$l \rightleftharpoons \alpha + \gamma \qquad e_2$

$\boxed{l \rightleftharpoons \alpha + \delta + \gamma} \qquad E_1$

$\underline{\alpha + \delta + \gamma}$

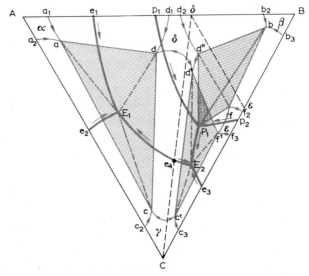

Fig. 210. Projection of the equilibria in the system of Fig. 209c on the concentration triangle.

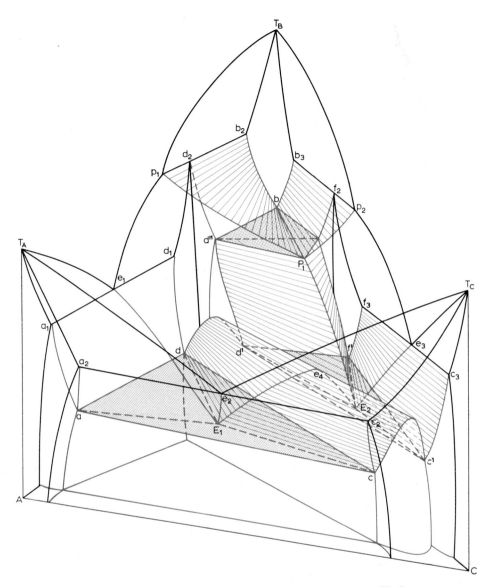

Fig. 211. Ternary space model corresponding to Fig. 210 (simplified).

four-phase planes shaded, but the decrease in solubility of the phases from the four-phase plane to room-temperature is ignored. Figure 210 should be compared with the simplified space model of Fig. 211. The δ phase region is given in Fig. 212a and the vertical section from C to δ in Fig. 212b. The latter emphasises the near quasi-binary nature of this section.

The reader is left to verify that the vertical sections $A\varepsilon$ and $\delta\varepsilon$ are not quasi-binary.

11.2.2. *Ternary intermediate phases*

The composition of a congruently-melting ternary intermediate phase falls within the area enclosed by the projection of the liquidus (and solidus) surfaces on the concentration triangle.

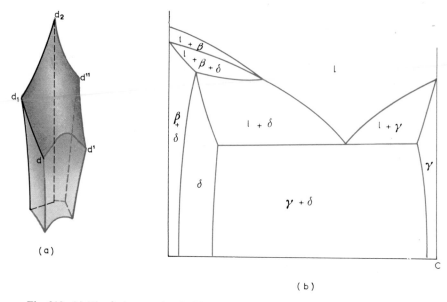

Fig. 212. (a) The δ phase region in Fig. 211; (b) a vertical section from C to δ.

The composition BiCuMg (Fig. 207) is within the projected area $e_1 E_1 e_2 E_3 e_3 E_2 e_1$ of the liquidus. An ordinate erected at the composition BiCuMg will pass through the common maximum point on the liquidus and solidus surfaces of the ternary phase. In other words, a congruently-melting ternary phase behaves as a pure metal (component) in that it freezes isothermally and its appearance is associated with a maximum on the liquidus/solidus surfaces.

An incongruently-melting intermediate phase is to be compared with its binary counterpart (Fig. 74) insofar as it is associated with a concealed maximum in the liquidus/solidus surfaces. The point representing the composition of a ternary incongruent phase (point 2 in Fig. 213) will lie outside the projected area of the liquidus/solidus surfaces.

If only one ternary intermediate phase is formed and all three binaries are eutectic, there are four cases to be distinguished. These can be classified according to whether one, two, three, or none of the sections from the ternary phase to the components are quasi-binary sections. The first case only will be considered. The space model and its projection are given in Fig. 213a–b. There are three invariant four-phase planes, corresponding to the equilibria:

$$P_1 \qquad l_{P_1} + \beta_b \rightleftharpoons \alpha_a + T_t$$
$$P_2 \qquad l_{P_2} + \beta_{b^1} \rightleftharpoons \gamma_{c^1} + T_{t^1}$$
$$E \qquad l_E \rightleftharpoons \alpha_{a^{11}} + \gamma_{c^{11}} + T_{t^{11}}$$

where T is a solid solution based on the ternary intermediate phase T. A quasi-binary section is assumed between T and β. This section is a peritectic type involving the reaction of liquid with β to form T:

$$l_1 + \beta_3 \rightleftharpoons T_2 .$$

The compositions of all three phases lie on one straight line 1 2 3 and the peritectic reaction occurs therefore at a maximum temperature. This point can be verified by considering the $l + \beta + T$ phase region which starts at the tie triangle $b t P_1$, ascends to the isotherm 1 2 3, degener-

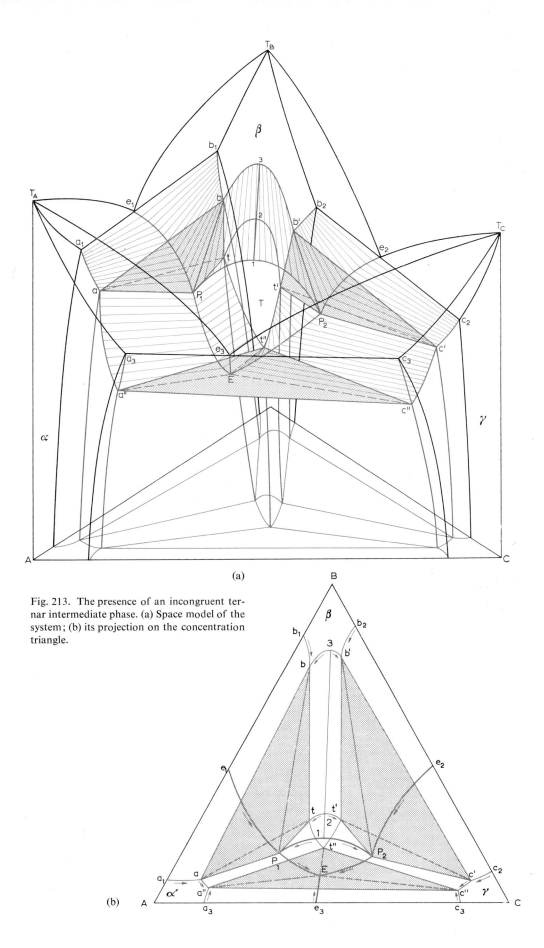

(a)

Fig. 213. The presence of an incongruent ternar intermediate phase. (a) Space model of the system; (b) its projection on the concentration triangle.

(b)

ates into a tie line and then descends to the tie triangle $b^1t^1P_2$. The region of existence of the T phase is illustrated in Fig. 214a; the $l+T$ phase region is given in Fig. 214b. The $\beta+T$ phase region is tunnel-shaped.

The change in solubility of the phases with fall in temperature from the four-phase planes is ignored in Fig. 213b. It usually happens that the original composition of the ternary phase (point 2 on line 1 2 3) is outside the area of existence of the T solid solution at room-temperature. Whether it is or not is dictated by the relative changes in solubility of the T phase on cooling from points t, t^1 and t^{11}. As will be seen from Fig. 213b, the T phase region $t_1t^1t^{11}$

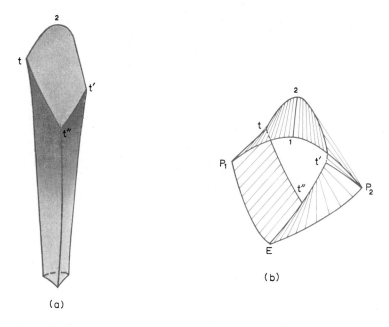

Fig. 214. Phase regions in the ternary system of Fig. 213a. (a) The T phase region; (b) the $l+T$ phase region.

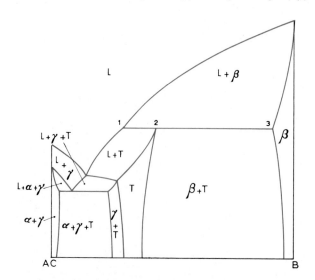

Fig. 215. The vertical section along tie line 1–2–3 of Fig. 213a.

TABLE 9

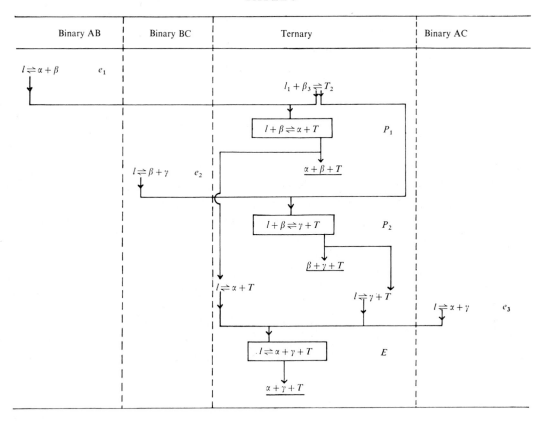

can slightly overlap the area of primary crystallisation of T, $P_1 1 P_2 E P_1$. The initial composition of the T phase (point 2) is always outside the area of primary crystallisation of T. Otherwise, T would be a congruently-melting ternary phase.

The vertical section which runs along tie line 1 2 3 illustrates the quasi-binary nature of the section in the region from B to point 1 (Fig. 215). A tabular representation of the system is given in Table 9.

An example of the occurrence of an incongruently-melting intermediate phase is found in the Al–Mg–Zn system* in which the ternary phase has a wide range of homogeneity.

* W. Köster and W. Dullenkopf, *Z. Metallk.*, **28** (1936) 367; also J. B. Clark, *Trans. Am. Soc. Metals* **53** (1961) 295.

CHAPTER 12

Ternary Phase Diagrams.
Liquid Immiscibility

Liquid immiscibility in one or more of the binary systems can lead to either three-phase or four-phase equilibria in the ternary system. Immiscibility can arise if either monotectic or syntectic reactions occur in the binary system; true ternary immiscibility is also possible.

As noted on p. 172, ternary three-phase equilibrium can involve monotectic and syntectic reactions leading to ternary systems in which there is a transition from a binary monotectic to another binary monotectic, or a transition from a monotectic to a syntectic, *etc.* The treatment of such systems is analogous to that adopted for the transitions eutectic/eutectic, peritectic/peritectic, eutectic/peritectic, *etc.* (Figs. 164, 169 and 170). A ternary system in which only one of the binaries exhibits liquid immiscibility has been referred to earlier (Fig. 172).

Liquid immiscibility leading to four-phase invariant equilibria remains to be considered. An interesting sequence of systems can be traced as a function of the number of binary systems which contain monotectic miscibility gaps. In each case the four-phase reaction is of the type:

$$l_1 \rightleftharpoons l_2 + \alpha + \gamma.$$

12.1. TWO BINARY SYSTEMS ARE MONOTECTIC

The AB and BC binaries are monotectics, the AC binary is eutectic. The ternary equilibria are represented in tabular form below and in the form of a projection on the concentration triangle in Fig. 216.

The miscibility gap in the AB binary stretches across the ternary system to meet the miscibility gap in the BC system. An example of this type of system is found in the Cd–Ga–Pb alloy system[*]. The four-phase monotectic reaction (Fig. 217) involves the disappearance of a liquid of composition M when forming another liquid of composition L_2 and two solid phases. The solid phases are virtually pure Cd and Pb. The monotectic plane is the triangle $Cd - L_2 - Pb$. From the monotectic plane at 234 °C the $L_2 + Cd + Pb$ equilibrium descends to the ternary eutectic plane at point E and 29.28 °C. On the assumption that the ternary eutectic liquid, E, behaves as an ideal solution use can be made of Van 't Hoff's relation to calculate the Ga content of the eutectic. Using the relation

$$\frac{L_{Ga}(T_0 - T)}{R T_0 T} = -\ln X_{Ga}$$

where L_{Ga} is the heat of fusion of Ga (1336 cal/g.-atom), T_0 is the m.p. of Ga (302.93 °K), T is the ternary eutectic temperature, R the gas constant, and X_{Ga} the Ga content of the ternary

---

[*] B. PREDEL, *Z. Metallk.*, **52** (1961) 507.

TABLE 10

Binary AB	Binary BC	Ternary	Binary AC
$l_1 \rightleftharpoons l_2 + \alpha$	$[l_1] \rightleftharpoons [l_2] + [\gamma]$		$(l_1) \rightleftharpoons (\alpha) + (\gamma)$
		$l_{1(M)} \rightleftharpoons \underline{l_2} + \underline{\alpha} + \underline{\gamma}$ M	
		$l_2 \rightleftharpoons \alpha + \gamma$	
$\langle l_2 \rangle \rightleftharpoons \langle \alpha \rangle + \langle \beta \rangle$	$\{l_2\} \rightleftharpoons \{\beta\} + \{\gamma\}$		
		$l_{2(E)} \rightleftharpoons \underline{\underline{\alpha}} + \underline{\underline{\beta}} + \underline{\underline{\gamma}}$ E	
		$\underline{\alpha + \beta + \gamma}$	

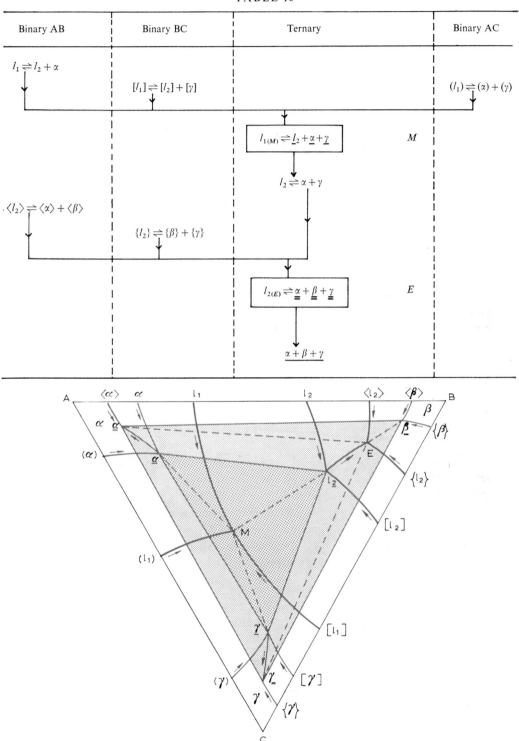

Fig. 216. Liquid immiscibility in ternary systems. Projection of the system when two binaries contain monotectics.

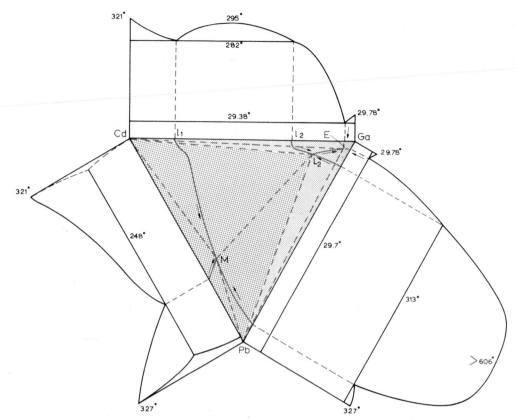

Fig. 217. The Cd–Ga–Pb system (schematic). (After B. PREDEL, *Z. Metallk.*, **52** (1961) 507; courtesy Dr. Riederer-Verlag GmbH.)

eutectic E, it is found that 99.63 %Ga is present in the ternary eutectic. In Fig. 217 point E has been drawn further into the triangle to clarify the equilibria.

One usually expects the curve connecting the critical points of the two binary miscibility gaps to flow smoothly from one binary to the other through the ternary system. In the Cd–Ga–Pb system the critical curve shows a minimum near the Cd–Ga binary. There is a corresponding minimum in the curves $l_1 M$ and $l_2 L_2$. This behaviour is taken by Predel to reflect a lower tendency to immiscibility in the ternary than in the two binary systems. The paper by Predel contains a series of very useful vertical sections, taken at constant %Ga across the diagram.

12.2. ONE BINARY SYSTEM IS MONOTECTIC

In this case (Fig. 218a–b) the miscibility gap does not extend from the binary AB to the binary BC. There is a four-phase reaction, $l_{1(M)} \rightleftharpoons \underline{l_2} + \alpha + \underline{\gamma}(M)$, which is preceded by the normal trio of three-phase equilibria. The first two are $l_1 \rightleftharpoons l_2 + \alpha$ from the AB binary and $(l_1) \rightleftharpoons (\alpha) + (\gamma)$ from the AC binary. The third presents the case of an upper critical point c_1 on the monovariant liquid curve $Mc_1 l_2$. At the critical temperature the liquid is in equilibrium with γ solid solution. The tie line $c_1 \gamma_1$ is a degenerate tie triangle. Below the critical temperature a three-phase equilibrium is observed between l_1 along curve $c_1 M$, l_2 along curve $c_1 \underline{l_2}$, and γ along curve $\gamma_1 \gamma$. A tabular representation can be expressed as in Table 11.

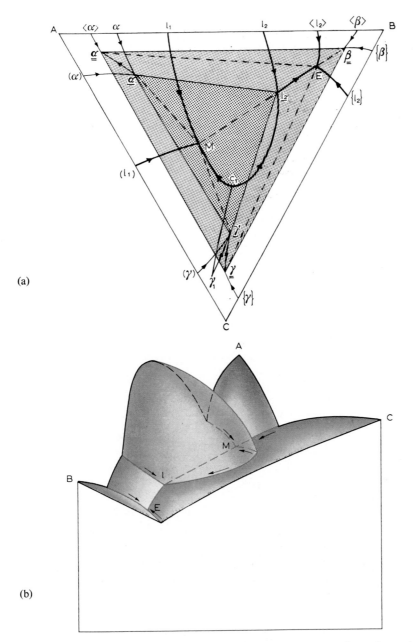

(a)

(b)

Fig. 218. Liquid immiscibility in ternary systems. (a) Projection of the system when only one binary is monotectic; (b) the corresponding liquidus surface.

A projection of the ternary system on the concentration triangle is given in Fig. 218a; the liquidus surface is illustrated in Fig. 218b. An example of this type of system is the Fe_3C–FeS–Fe partial system of the C–Fe–S ternary*. The quasi-binary system Fe_3C–FeS is monotectic and the Fe–Fe_3C and Fe–FeS systems are taken to be simple eutectics for this purpose.

* R. Vogel, *Die Heterogenen Gleichgewichte*, 2nd edn., Akademische Verlagsgesellschaft Geest & Portig K.-G., Leipzig, 1959, p. 608.

TABLE 11

Binary AB	Binary BC	Ternary	Binary AC
		$l_{c_1} \rightleftharpoons \gamma_1$	
		$\downarrow$	
		$l_1 + l_2 + \gamma$	
$l_1 \rightleftharpoons l_2 + \alpha$			$(l_1) \rightleftharpoons (\alpha) + (\gamma)$
$\downarrow$			$\downarrow$
		$\boxed{l_{1(M)} \rightleftharpoons \underline{l_2} + \underline{\alpha} + \underline{\gamma}}$ M	
		$\downarrow$	
		$l_2 \rightleftharpoons \alpha + \gamma$	
$\langle l_2 \rangle \rightleftharpoons \langle \alpha \rangle + \langle \beta \rangle$			
$\downarrow$	$\{l_2\} \rightleftharpoons \{\beta\} + \{\gamma\}$	$\downarrow$	
	$\downarrow$		
		$\boxed{l_{2(E)} \rightleftharpoons \underline{\underline{\alpha}} + \underline{\underline{\beta}} + \underline{\underline{\gamma}}}$ E	
		$\downarrow$	
		$\underline{\alpha + \beta + \gamma}$	

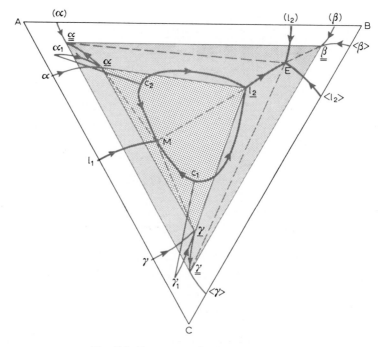

Fig. 219. True ternary liquid immiscibility.

TABLE 12

Binary AB	Binary BC	Ternary	Binary AC

The diagram contents, reading through the ternary reactions:

$$l_{c_2} \rightleftharpoons \alpha_1$$

$$l_1 + l_2 + \alpha$$

$$l_{c_1} \rightleftharpoons \gamma_1$$

$$l_1 + l_2 + \gamma$$

$$l_1 \rightleftharpoons \alpha + \gamma$$

$$l_{1(M)} \rightleftharpoons \underline{l_2} + \underline{\alpha} + \underline{\gamma} \qquad M$$

$$l_2 \rightleftharpoons \alpha + \gamma$$

$$(l_2) \rightleftharpoons (\alpha) + (\beta)$$

$$\langle l_2 \rangle \rightleftharpoons \langle \beta \rangle + \langle \gamma \rangle$$

$$l_{2(E)} \rightleftharpoons \underline{\alpha} + \underline{\beta} + \underline{\gamma} \qquad E$$

$$\underline{\alpha + \beta + \gamma}$$

12.3. NONE OF THE BINARIES CONTAIN LIQUID MISCIBILITY GAPS BUT TRUE TERNARY LIQUID IMMISCIBILITY APPEARS

Figure 218a was derived from Fig. 216 by reducing the number of binary liquid miscibility gaps from two to one. If this is taken a step further and all three binaries are assumed to be eutectics, a completely enclosed ternary liquid miscibility gap results. The projection of the ternary system is given in Fig. 219 and in Table 12.

The ternary invariant reaction at M is preceded by the monovariant eutectic equilibrium $l_1 \rightleftharpoons \alpha + \gamma$, and by two monovariant equilibria which originate at critical points c_1 and c_2. The $l_1 + l_2$ phase region has the shape of a sector of a sphere, with Ml_2 the diameter of the sphere.

Ternary miscibility gaps appear in a number of alloy systems. Reference should be made to the work of R. Vogel and colleagues with respect to the Cu–Fe–P*, Fe–P–Sb** and Fe–Sb–Si*** systems.

* R. VOGEL AND I. BERAK, *Arch. Eisenhüttenw.*, **21** (1950) 327.
** R. VOGEL AND D. HORSTMANN, *Arch. Eisenhüttenw.*, **23** (1952) 127.
*** R. VOGEL AND G. ZWINGMANN, *Arch. Eisenhüttenw.*, **28** (1957) 591.

CHAPTER 13

Ternary Phase Diagrams.
Four-Phase Equilibrium Involving Allotropy of One Component

In the transition from a binary diagram of the closed γ type to one of the expanded γ type (Fig. 220a–b) a four-phase equilibrium will appear. It is assumed that the BC binary shows a complete series of solid solutions. This type of ternary is of importance in the metallurgy of low alloy steels, since Fig. 220a is recognisable as the Fe–Fe$_3$C diagram and Fig. 220b as the

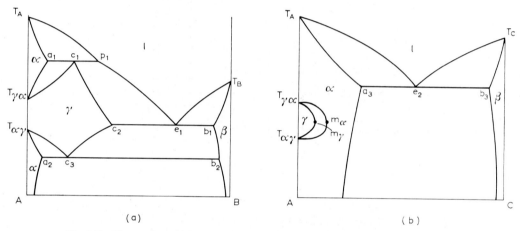

Fig. 220. Binary phase diagram (a) with an expanded γ field, (b) with closed γ field.

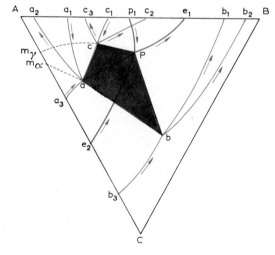

Fig. 221. Projection of the ternary system involving a transition from a closed γ field to an expanded γ field.

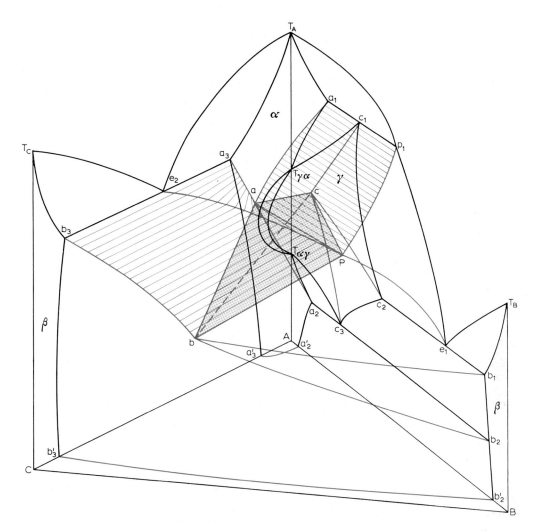

Fig. 222. Ternary space model corresponding to Fig. 221. For the sake of clarity the BC binary system is not included.

type of diagram produced by ferrite-forming elements such as Cr, Mo, Si and W. In the ternary there appears a four-phase equilibrium, $l + \alpha \rightleftharpoons \beta + \gamma$. The four-phase plane, *Pbac*, is shaded in the projection of the system (Fig. 221) and in the space model (Fig. 222). A tabular representation of the ternary is given in Table 13.

The monovariant $l + \alpha \rightleftharpoons \gamma$ equilibrium originates at the peritectic horizontal $a_1 c_1 p_1$ on the AB binary and descends to the four-phase plane as the triangle *acP*; the monovariant $l \rightleftharpoons \alpha + \beta$ equilibrium originates at the eutectic horizontal $a_3 e_2 b_3$ on the AC binary and descends to the four-phase plane as the triangle *aPb*. As a result of the ternary quasi-peritectic reaction, $l_P + \alpha_a \rightleftharpoons \beta_b + \gamma_c$, two monovariant equilibria descend from the four-phase plane to lower temperatures. These are the $l \rightleftharpoons \beta + \gamma$ equilibrium, descending from triangle *cPb* to the eutectic horizontal $c_2 e_1 b_1$ on the AB binary, and the $\gamma \rightleftharpoons \alpha + \beta$ equilibrium descending from triangle *acb* to the eutectoid horizontal $a_2 c_3 b_2$ on the AB binary. As will be evident from an inspection of Figs. 221 and 222, the $\gamma \rightleftharpoons \alpha + \beta$ reaction leads to an $\alpha\beta\gamma$ phase region with a peculiar

TABLE 13

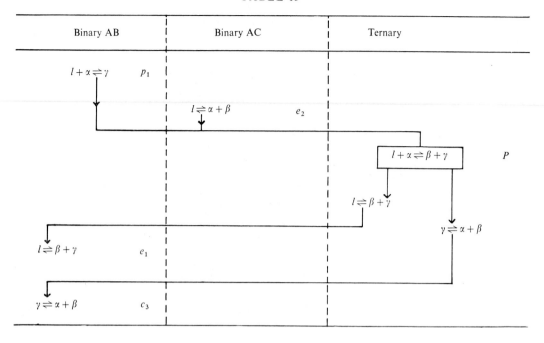

Binary AB	Binary AC	Ternary
$l + \alpha \rightleftharpoons \gamma$ p_1		
	$l \rightleftharpoons \alpha + \beta$ e_2	
		$l + \alpha \rightleftharpoons \beta + \gamma$ P
		$l \rightleftharpoons \beta + \gamma$
		$\gamma \rightleftharpoons \alpha + \beta$
$l \rightleftharpoons \beta + \gamma$ e_1		
$\gamma \rightleftharpoons \alpha + \beta$ c_3		

shape. The $\alpha\gamma$ ruled surface descends from the tie line ac to the tie line a_2c_3 in a straightforward manner; the $\alpha\beta$ ruled surface skews round from ab to a_2b_2 and the $\beta\gamma$ ruled surface also skews round from bc to b_2c_3.

The only two-phase region which presents any difficulty in perception is the $\alpha + \gamma$ phase region. This region originates at the tie line a_1c_1 on the peritectic horizontal $a_1c_1p_1$ and ends at the tie line a_2c_3 on the eutectoid horizontal $a_2c_3b_2$. The part of the phase region facing the reader in Fig. 222 is formed by the ruled surfaces $a_1c_1caa_1$ and acc_3a_2a; the other part of the region is constrained to follow the $\alpha + \gamma$ phase region in the binary AB and AC systems. Beginning at a_1c_1, it follows the $\alpha + \gamma$ phase region on the AB binary until the $\gamma \rightleftharpoons \alpha$ transition temperature, $T_{\gamma\alpha}$ is reached; it then turns through $60°$ and follows the loop-shaped $\alpha + \gamma$ region on the AC binary until the $\alpha \rightleftharpoons \gamma$ transition temperature, $T_{\alpha\gamma}$, is reached. It finally turns back through $60°$ on to the AB binary to follow the $\alpha + \gamma$ region down to a_2c_3. The $\alpha + \gamma$ phase

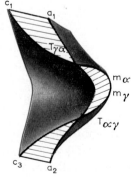

Fig. 223. The $\alpha + \gamma$ phase region of Fig. 221.

region in the ternary system has therefore the rough shape of a thick, bulged letter C (Fig. 223). A good way of verifying the shape is to model the phase region from plasticene.

The dotted lines from c to m_γ and from a to m_α (Fig. 221) are intended to remind the reader that the $\alpha+\gamma$ phase region starts on the AB binary, moves to the AC binary and then back to the AB binary again as the temperature falls. It is not meant to suggest that the four-phase plane exists at the same temperature as the tie line $m_\gamma m_\alpha$.

Vertical sections

The transition from the closed γ field to the expanded γ field is well illustrated by taking a series of vertical sections through the ternary diagram (Figs. 224 and 225). Each vertical section passes through corner A. The relative positions of the sections is shown in Fig. 224. Where a section crosses a line or a curve in the ternary the intersection is noted by a number. This helps in the construction of the sections in Fig. 225.

Section A–I (Fig. 225a) intersects the α liquidus surface $T_A e_2 P p_1 T_A$ over the curve $T_A 2$ and the β liquidus surface $T_C e_2 P e_1 T_B T_C$ over the curve which ascends from point 2 to the ordinate on the right. Point 2 lies on the monovariant liquid curve $e_2 P$. The section intersects the $l\alpha\beta$ phase region to give the triangular area 1–2–3, point 1 lying on curve $a_3 a$ and point 3 on curve $b_3 b$. The solubility curves from points 1 and 3 to room-temperature are the traces of

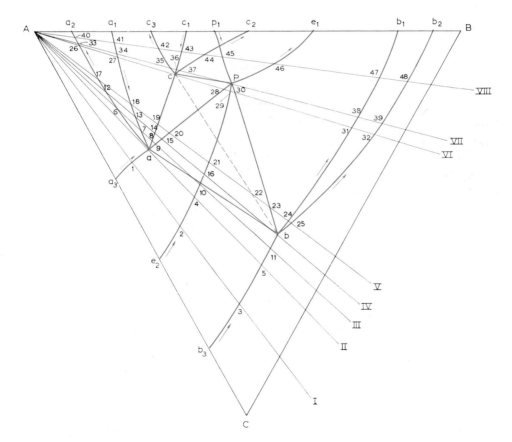

Fig. 224. Location of vertical sections through Fig. 221.

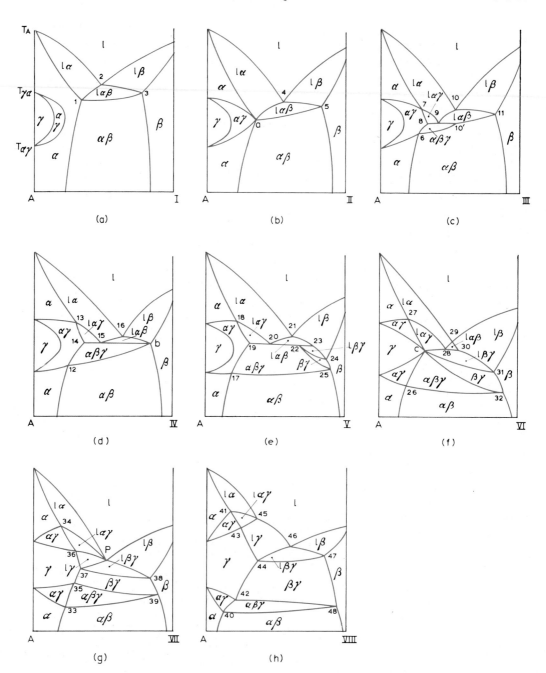

Fig. 225. Vertical sections through the space model of Fig. 221.

the section over the α solubility surface $a_3aa_2a_2^1a_3^1a_3$ and the β solubility surface $b_3bb_2b_2^1b_3^1b_3$. By reference to Fig. 223 it can be seen that the vertical section A–I cuts the $\alpha + \gamma$ phase region to give a similar-shaped γ loop as that shown in the binary AC system. The only essential difference between Fig. 225a and the binary AC system (Fig. 220b) is that the binary horizontal $a_3e_2b_3$ is replaced by a three-phase $l\alpha\beta$ area (1–2–3) in the ternary section.

Section A–II is very similar to A–I but, as it passes through point a on the four-phase plane, the $\alpha + \gamma$ phase region (denoted $\alpha\gamma$ for short in Fig. 225b) must just stretch to point a. In considering the shape of the $\alpha\gamma$ region it must be remembered that the $\alpha\gamma$ region in the ternary sweeps down from a_1c_1 to ac and back again to a_2c_3. The α surface of the $\alpha\gamma$ region therefore starts at a_1, moves down to a and then back to a_2. Section A–II will cut the α surface of the $\alpha\gamma$ region from $T_{\gamma\alpha}$ down to a and from a down to $T_{\alpha\gamma}$. The γ surface of the $\alpha\gamma$ region does not reach point a so there is still the characteristic γ loop present.

Section A–III cuts across the four-phase plane near the a corner. It meets the four-phase plane at point 8 and extends to point 10. The line 8–9–10 is therefore horizontal in Fig. 225c. It should be noted that point 10 is both a point on the four-phase plane and also a point on the monovariant liquid curve e_2P but that the latter intersection is at a higher temperature than the former. The section cuts the $l\alpha\gamma$ region to produce the area 7–8–9; curve 7–8 is on the $\alpha\gamma$ ruled surfaces $a_1acc_1a_1$, curve 7–9 on the $l\alpha$ ruled surface $a_1aPp_1a_1$. Similarly, the section cuts the $l\alpha\beta$ region over the $l\alpha$ ruled surface $a_3aPe_2a_3$ to give 9–10, over the $l\beta$ ruled surface $e_2Pbb_3e_2$ to give curve 10–11, and over the $\alpha\beta$ ruled surface $b_3baa_3b_3$ to give curve 11–10^1. Finally, the section cuts the $\alpha\beta\gamma$ phase region below the four-phase plane to give curve 8–6 over the $\alpha\gamma$ ruled surface acc_3a_2a and curve 10^1–6 over the $\alpha\beta$ ruled surface abb_2a_2a.

Section A–IV (Fig. 225d) is a similar section. The two main points of difference result from the section passing through point b on the four-phase plane. The horizontal line now extends from 14 to b and the $\alpha\beta\gamma$ phase region has been extended across the region to the point b. The next section, A–V (Fig. 225e), only differs in respect to the phase regions below the four-phase plane. From the edge of the four-phase plane, at point 23, the section cuts over the $l\beta$ ruled surface bPe_1b_1b along curve 23–24. From 24 the section intersects the $\beta\gamma$ ruled surface bcc_2b_1b along curve 24–22; point 22 is on the tie line bc of the four-phase plane. The area 22–23–24 encloses the $l\beta\gamma$ phase region. From point 22 the section cuts over the $\beta\gamma$ ruled surface bcc_3b_2b to give curve 22–25. The $\beta\gamma$ phase region, 22–25–24, is completed by considering intersection of the section with the β solubility surface bb_1b_2b along the curve ascending from 25 to 24.

When the vertical section passes through point c, as section A–VI, the γ surface of the $\alpha\gamma$ phase region now reaches the four-phase plane from above and below (Fig. 225f) (*i.e.* from $T_{\gamma\alpha}$ and $T_{\alpha\gamma}$). The $l\beta\gamma$ and $\beta\gamma$ phase regions have moved across the diagram to meet at point c. Sections nearer the AB binary will show a separation of the $\alpha\gamma$ phase region into an upper and lower phase region. Section A–VII illustrates this event. The $\beta\gamma$ phase region now forms a broad band across the diagram between the β and γ phase fields and the $\alpha\beta\gamma$ region has shrunk downwards. In the upper half of Fig. 225g a new phase region has appeared—$l\gamma$. In sections closer to the AB binary (section A–VIII, Fig. 225h) the $l\gamma$ region has grown and the diagram is recognisable as the ternary equivalent of Fig. 220a. In place of the binary horizontals $a_1c_1p_1$, $c_2e_1b_1$ and $a_2c_3b_2$ there are the ternary three-phase regions $l\alpha\gamma$ (41–43–45), $l\beta\gamma$ (44–46–47) and $\alpha\beta\gamma$ (40–42–48).

By taking this series of vertical sections we have seen how the closed γ binary (Fig. 220b) is transformed to the expanded γ binary (Fig. 220a) through the ternary.

CHAPTER 14

The Association of Phase Regions

14.1. LAW OF ADJOINING PHASE REGIONS

In the construction of phase diagrams the Phase Rule imposes certain restrictions on the disposition of the phase regions. With binary systems it is easily seen that a single phase region is separated by a line from a two-phase region and that no two single-phase regions adjoin each other through a line. Masing* was the first to attempt to define rules for adjoining phase regions in ternary systems. He noted that "a state space can ordinarily be bounded by another state space only if the number of phases in the second space is one less or one greater than that in the first space considered". A single-phase space such as the liquid, α, β, etc., phase regions is separated from a two-phase region (*e.g.* $l+\alpha$, $l+\beta$ phase regions) by a surface in a space model. A two-phase region is separated from a three-phase region by a surface—a ruled surface. Finally, a single-phase region meets a three-phase region along a curve.

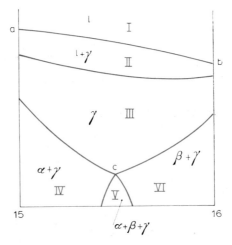

Fig. 226. Application of the law of adjoining phase regions to the vertical section of Fig. 178h.

The most useful rule on the association of phase regions is due to Palatnik and Landau**. They derived a general relationship for the dimension, R_1, of the boundary between neighbouring phase regions:

$$R_1 = R - D^- - D^+ \geq 0.$$

The relation is referred to as the Law of Adjoining Phase Regions. R is the dimension of the phase diagram or section of the diagram (vertical or isothermal) and D^- and D^+ represent,

---

 * G. MASING, *Ternary Systems*, (translated by B. A. Rogers), Reinhold, New York, 1944, p. 34.
 ** L. S. PALATNIK AND A. I. LANDAU, *Zh. Fiz. Khim.*, **29** (1955) 1784, 2054; **30** (1956) 2399.

respectively, the number of phases that disappear and the number that appear in a transition from one phase region to the other.

Example 1. To illustrate the application of the law of adjoining phase regions, the vertical section (Fig. 178h) is re-drawn in Fig. 226 with each phase region denoted by a number. The vertical section is two-dimensional and so $R = 2$. Table 14 summarises the method for establishing the dimension of the boundaries between neighbouring phase regions.

TABLE 14

Transition	I	II	II	III	III	IV	III	IV	III	V	IV	V	VI	V
	↓	↓	↓	↓	↓	↓	↓	↓	↓	↓	↓	↓	↓	↓
from:	II	I	III	II	IV	III	VI	III	V	III	V	IV	V	VI
R	2	2	2	2	2	2	2	2	2	2	2	2	2	2
D^-	0	1	1	0	0	1	0	1	0	2	0	1	0	1
D^+	1	0	0	1	1	0	1	0	2	0	1	0	1	0
R_1	1	1	1	1	1	1	1	1	0	0	1	1	1	1
Corresponding geometrical element	line		line		line		line		point		line		line	

Taking transition I → II, we are going from the liquid phase region to the $l+\gamma$ phase region. In the transition $l \to l+\gamma$ no phases disappear but one new phase appears: γ. On the other hand, in the reverse transition from II → I, $l+\gamma \to l$, the γ phase disappears but no new phase appears at the end of the transition. Irrespective of the direction of the transition, I → II or II → I, the resulting value of R_1, the dimension of the boundary between the phase regions, is the same. In this case the boundary is one-dimensional and is represented by the line ab.

For the transition IV → V, $\alpha+\gamma \to \alpha+\beta+\gamma$, no phase disappears ($D^- = 0$) but one phase appears ($D^+ = 1$) in the transition—the β phase. Again $R_1 = 1$ and the boundary between the $\alpha+\gamma$ and $\alpha+\beta+\gamma$ phase regions is the line cd.

For the transition V → III, $\alpha+\beta+\gamma \to \gamma$, the α and β phases disappear in crossing from phase region V to III. Since no new phase appears $R_1 = R - D^- - D^+ = 2-2-0 = 0$. The phase boundary has zero dimensions; it is represented by point c.

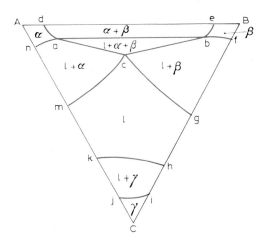

Fig. 227. Application of the law of adjoining phase regions to the isothermal section of Fig. 176c.

Example 2. The ternary space model illustrated in Fig. 164a and in an exploded form in Fig. 165 will be taken as an example of a three-dimensional phase diagram. Transitions from a single-phase region, a two-phase region and a three-phase region will be considered: *i.e.* $\beta \to l+\beta$, $l+\alpha+\beta$, and $\alpha+\beta$; $l+\beta \to l$, β, $\alpha+\beta$, and $l+\alpha+\beta$; $l+\alpha+\beta \to l$, α, β, $l+\alpha$, $l+\beta$, and $\alpha+\beta$.

TABLE 15

Transition from:	β			$l+\beta$				$l+\alpha+\beta$					
	$l+\beta$	$l+\alpha+\beta$	$\alpha+\beta$	l	β	$\alpha+\beta$	$l+\alpha+\beta$	l	α	β	$l+\alpha$	$l+\beta$	$\alpha+\beta$
R	3	3	3	3	3	3	3	3	3	3	3	3	3
D^-	0	0	0	1	1	1	0	2	2	2	1	1	1
D^+	1	2	1	0	0	1	1	0	0	0	0	0	0
R_1	2	1	2	2	2	1	2	1	1	1	2	2	2
Corresponding geometrical element	a	b	c	d	e	f	g	h	i	j	k	l	m

a – surface ($T_B bb_1 T_B$), b – line (bb_1), c – surface ($bb_1 d_1 db$), d – surface ($T_B ee_1 T_B$), e – surface ($T_B bb_1 T_B$), f – line (bb_1), g – surface ($bb_1 e_1 eb$), h – line (ee_1), i – line (aa_1), j – line (bb_1), k – surface ($aee_1 a_1 a$), l – surface ($bb_1 e_1 eb$), m – surface ($abb_1 a_1 a$).

It is apparent that a single-phase region adjoins a two-phase region over a surface, and a three-phase region along a line. A two-phase region adjoins a three-phase region over a surface (a ruled surface) and a two-phase region along a line. A surface or a line separates any two neighbouring phase regions and four phase regions meet along a line (*cf.* Fig. 165).

Example 3. The application of the law of adjoining phase regions to isothermal sections can be considered with reference to Fig. 227 (*cf.* Fig. 176c.) The following transitions from a single-phase region to its neighbours are possible:

$$\alpha \to \alpha+\beta, \quad l+\alpha, \quad l+\alpha+\beta$$
$$\beta \to \alpha+\beta, \quad l+\beta, \quad l+\alpha+\beta$$
$$l \to l+\alpha, \quad l+\beta, \quad l+\gamma, \quad l+\alpha+\beta$$
$$\gamma \to l+\gamma$$

Other transitions, *e.g.* $\alpha+\beta \to \alpha$, *etc.*, are possible, but their consideration is superfluous.

TABLE 16

Transition from:	α			β			l				γ
	$\alpha+\beta$	$l+\alpha$	$l+\alpha+\beta$	$\alpha+\beta$	$l+\beta$	$l+\alpha+\beta$	$l+\alpha$	$l+\beta$	$l+\gamma$	$l+\alpha+\beta$	$l+\gamma$
R	2	2	2	2	2	2	2	2	2	2	2
D^-	0	0	0	0	0	0	0	0	0	0	0
D^+	1	1	2	1	1	2	1	1	1	2	1
R_1	1	1	0	1	1	0	1	1	1	0	1
Corresponding geometrical element	line (da)	line (na)	point (a)	line (eb)	line (bf)	point (b)	line (mc)	line (gc)	line (kh)	point (c)	line (ji)

14.2. DEGENERATE PHASE REGIONS

The examples used above to illustrate the applicability of the law of adjoining phase regions to space models and their vertical and isothermal sections were chosen so that no invariant reaction isotherm or four-phase plane was included. An invariant reaction in a phase diagram or a section through a phase diagram is a degenerate phase region. Such a phase region can be defined as one which has a dimension less than that of the phase diagram or its section. Degenerate phase regions can be considered as phase regions with one or more degenerate dimensions, *i.e.* dimensions that have vanished. In considering phase diagrams or sections containing degenerate phase regions, it is necessary to replace the missing dimensions before applying the law of adjoining phase regions.

To illustrate this point, consider the simple case of the melting of a pure component A (Fig. 228a). We know experimentally that such a process occurs at a unique temperature, the melting point T_A. At T_A both liquid and solid are in equilibrium and the Phase Rule tells us that this equilibrium is invariant ($p+f = c+1$; $2+f = 1+1$). Applying the law of adjoining phase regions to the transition from the solid phase region to the liquid phase region (Fig. 228a), we find that the boundary between the two phase regions has a dimension of -1 ($R = 1$, $D^- = 1$, $D^+ = 1$; therefore $R = -1$). To overcome this situation, one regards the point T_A as a degenerate (liquid + solid) phase region and one replaces the missing dimension to give the diagram shown in Fig. 228b. This is now a topologically correct diagram which obeys the law of adjoining phase regions but not the Phase Rule. The replacement of missing dimensions in degenerate phase regions is a very useful method for checking the construction of phase diagrams, provided that it is remembered that such a replacement can lead to violations of the Phase Rule.

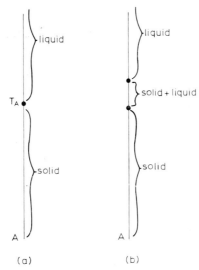

Fig. 228. Illustration of a degenerate phase region. (a) The melting of pure A; (b) the melting of pure A when point T_A is regarded as a degenerate phase region and replaced by a "solid + liquid" phase region.

Degenerate phase regions in space models of phase diagrams and in their sections can be dealt with in a similar manner by replacing the missing dimensions. A binary eutectic diagram (Fig. 229a) has an invariant reaction isothermal *aeb*. The tie line *ae* separates the $l+\alpha$ phase region from the $\alpha+\beta$ phase region, and the tie line *eb* the $l+\beta$ from the $\alpha+\beta$ phase region. The eutectic horizontal *aeb* represents the conditions of temperature and composition for which the three phases l, α and β are in equilibrium. It is thus a three-phase region with only one dimension. In Fig. 229b the degeneracy of the phase region *aeb* has been eliminated by making

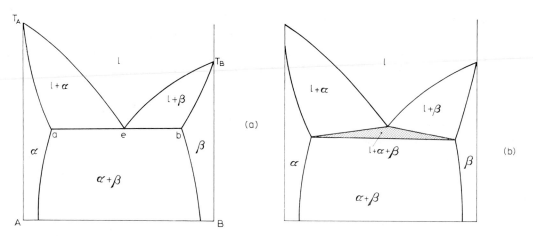

Fig. 229. Illustration of degenerate phase regions. (a) The eutectic phase diagram; (b) corresponding diagram allowing for degeneration of the phase regions.

it a two-dimensional phase region. In doing so, one is emphasizing the three-phase nature of the eutectic reaction and producing a diagram which complies with the law of adjoining phase regions. It will be noted that the liquidus and solidus curves do not meet at T_A, in conformity with Fig. 228b.

Sections through invariant four-phase planes in ternary systems can be considered in a similar manner. The section 9–10 (Fig. 178e) through the ternary eutectic four-phase plane would appear as in Fig. 230 if the degenerate phase region $l+\alpha+\beta+\gamma$ were eliminated. In the ternary space model (Fig. 173a) this implies that the four-phase plane abc is converted into a

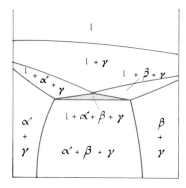

Fig. 230. Degeneration phase regions in vertical sections through ternary space models (Fig. 178e).

flat tetrahedron with base abc and upper vertex E close to the base. If one studies Fig. 173a, it is clear that a single-phase region meets a four-phase region (the degenerate four-phase plane or the four-phase tetrahedron) at a point. Such a point could be a, where the α phase region meets the $l+\alpha+\beta+\gamma$ phase region. A two-phase region meets a four-phase region along a tie line. In Fig. 175a the $l+\beta$ phase region meets the $l+\alpha+\beta+\gamma$ phase region along the tie line bE. Finally, a three-phase region such as $l+\alpha+\beta$ (Fig. 174), meets a four-phase region over a surface—the tie triangle abE.

In three-dimensional representations of ternary systems the junction of various phase regions can be summarised as follows:

(1) a single-phase region with a two-phase region over a surface,
(2) a single-phase region with a three-phase region along a line (non-isothermal),
(3) a single-phase region with a four-phase region at a point,
(4) a two-phase region with a three-phase region over a ruled surface,
(5) a two-phase region with a four-phase region along a tie line,
(6) a three-phase region with a four-phase region over a tie triangle,
(7) a surface separates two neighbouring phase regions,
(8) four neighbouring phase regions meet along a common line,
(9) six neighbouring phase regions meet at a common point.

14.3. TWO-DIMENSIONAL SECTIONS OF PHASE DIAGRAMS

The boundary between adjoining phase regions in a two-dimensional phase diagram or a two-dimensional section of a phase diagram can be either a line or a point, corresponding to $R_1 = 1$ and $R_1 = 0$ respectively. That the boundary cannot have the same dimension as the phase diagram or its section is self-evident. If it had, $R = R_1 = 2$, and $D^- + D^+ = 0$, or $D^- = D^+ = 0$. Since no phases disappear or are formed in going from one phase region to the other, the phase regions are identical. In other words, in an R-dimensional phase diagram or section of a phase diagram the boundary between adjoining phase regions cannot have a dimension greater than $R-1$, i.e. $R_1 \leqslant R-1$. This relation will be recognised as the general definition of a boundary.

(a) $R = 2$; $R_1 = 1$

This case is represented in Fig. 231. The boundary between phase regions I and II is a line. The number of phases in region I is $\lambda(\alpha_1 + \alpha_2 + \ldots \ldots + \alpha_\lambda)$. To satisfy the law of adjoining phase regions the number of phases in region II must be $\lambda + 1(\alpha_1 + \alpha_2 + \ldots \ldots + \alpha_\lambda + \alpha_{\lambda+1})$, it being assumed that region I has the lower number of phases. Comparison with the binary phase diagrams and two-dimensional sections of ternary phase diagrams in the preceding chapters illustrates the validity of the concept that a line separates phase regions containing λ and $\lambda+1$ phases (α from $\alpha+\beta$, Fig. 220a; $\alpha+\gamma$ from $\alpha+\beta+\gamma$, Fig. 178e; and $l+\alpha+\gamma$ from $l+\alpha+\beta+\gamma$, Fig. 230). As stressed previously, the missing dimensions have to be added to degenerate phase regions to allow application of the law.

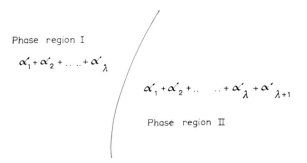

Fig. 231. Phase distribution in a two-dimensional diagram when the boundary between adjoining phase regions is one-dimensional.

(b) $R = 2^3$; $R_1 = 0$

At first sight it might seem possible for three boundary lines to meet at a point in a two-dimensional diagram (Fig. 232). We will apply the law of adjoining phase regions to determine the phase compositions of phase region III. Start with the transition from region I to region III. Three compositions are then possible for region III:

(1) $\alpha_1 + \alpha_2 + \ldots$ $\ldots + \alpha_\lambda + \alpha_{\lambda+1}$

(2) $\alpha_1 + \alpha_2 + \ldots$ $\ldots + \alpha_\lambda + \alpha_{\lambda+2}$

(3) $\alpha_1 + \alpha_2 + \ldots$ $\ldots + \alpha_{i-1} + \alpha_{i+1} + \ldots$ $\ldots + \alpha_\lambda$ (the absence of one of the phases, α_i, found in region I).

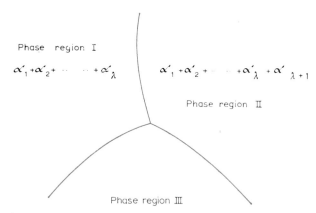

Fig. 232. Impossibility of three boundary lines meeting at a point in a two-dimensional diagram.

If we now consider the transition from region III to region II it is evident that none of the three possible phase compositions for region III satisfy the law of adjoining phase regions. At least four lines must meet at a point in a two-dimensional diagram. In general, only four lines meet at a point in a two-dimension diagram. The exceptions to this rule are given below.

Consider the situation where five lines meet at a point (Fig. 233a). Each phase region is separated from its neighbours by lines and from all other phase regions by point O. It is assumed that all phase regions have different compositions and that the region with the least number of phases is region I with λ phases. Then regions II and III will have compositions $\alpha_1 + \alpha_2 + \ldots$ $\ldots + \alpha_\lambda + \alpha_{\lambda+1}$ and $\alpha_1 + \alpha_2 + \ldots \ldots + \alpha_\lambda + \alpha_{\lambda+2}$ respectively. Applying the law of adjoining phase regions to a transition from region II to region IV there are four possible compositions for region IV:

(1) $\alpha_1 + \alpha_2 + \ldots$ $\ldots + \alpha_\lambda + \alpha_{\lambda+1} + \alpha_{\lambda+2}$

(2) $\alpha_1 + \alpha_2 + \ldots$ $\ldots + \alpha_\lambda + \alpha_{\lambda+1} + \alpha_{\lambda+3}$

(3) $\alpha_1 + \alpha_2 + \ldots$ $\ldots + \alpha_{i-1} + \alpha_{i+1} + \ldots$ $\ldots + \alpha_\lambda + \alpha_{\lambda+1}$

(4) $\alpha_1 + \alpha_2 + \ldots$ $\ldots + \alpha_\lambda$.

The fourth variant is not permissible since the phase regions must have different compositions. The second and third do not satisfy the law for transition from region IV to region III through point O since $R_1 = -1$. The first variant only satisfies the law for a transition from region IV to region III if the boundary between regions III and IV is a line. Phase region V cannot exist; only four lines may meet at a point in two-dimensional diagrams (Fig. 233c). As an example, consider the phase regions meeting at point c in Fig. 226. The disposition of the regions has been redrawn in Fig. 233b which should be compared with the generalised state of affairs represented in Fig. 233c. In Fig. 233b the phase region with the least phases is the γ phase region. The γ region is equivalent to the $\alpha_1 + \alpha_2 + \ldots \ldots + \alpha_\lambda$ region in Fig. 233c. The $\alpha_1 + \alpha_2 + + \ldots \ldots + \alpha_\lambda + \alpha_{\lambda+1}$ region (II) contains one more phase than region I—$\alpha_{\lambda+1}$, equivalent to the α phase in Fig. 233b. Similarly, the region III contains a phase $\alpha_{\lambda+2}$ equivalent to the β phase. Both the $\alpha_{\lambda+1}$ and $\alpha_{\lambda+2}$ are present in region IV in addition to the basic phase $\alpha_1 + \alpha_2 + + \ldots \ldots + \alpha_\lambda$. In similar fashion both α and β are present with the γ phase in the corresponding region in Fig. 233b.

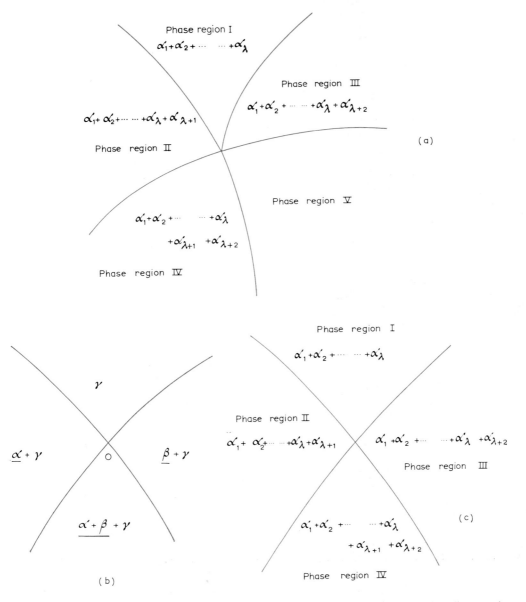

Fig. 233. Boundary lines meeting at a point in a two-dimensional diagram. (a) Impossibility of five lines meeting at a point; (b) distribution of phase regions when four lines meet at a point; (c) only four lines may meet at a point.

That there are exceptions to the rule that four lines meet at a point in a two-dimensional diagram is evident from an examination of Fig. 178b and f. In each case six lines meet at a central point. It will be noted, however, that in both cases the section passes through an invariant point—E and c respectively. Palatnik and Landau call such sections nodal or non-regular sections. Only regular sections obey the law of adjoining phase regions completely.

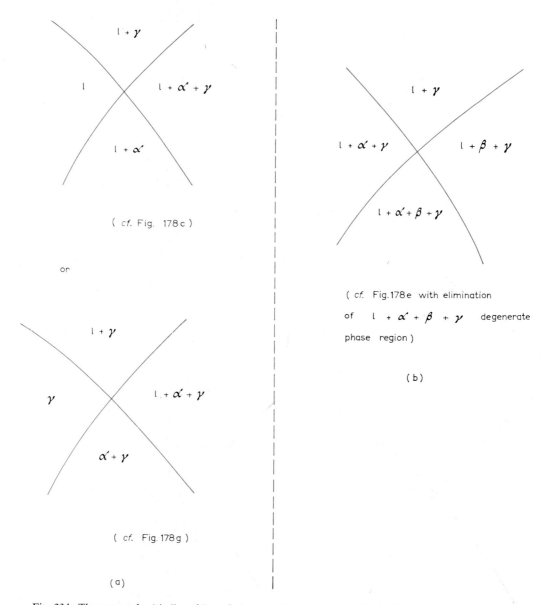

Fig. 234. The cross rule, (a) disposition of phase regions when one region is $l+\gamma$, (b) alternative disposition of phase regions.

14.4. THE CROSS RULE

The phase distribution in the four phase regions of Fig. 233b or c is determined in accordance with the cross rule*. Of the four phase regions one will have the least number of phases. In Fig. 233b this would be the γ phase region. The neighbouring phase regions each have an additional phase—the $\alpha + \gamma$ and $\beta + \gamma$ phase regions. The fourth phase region adjoins the phase region with the least number of phases through point 0. It is the phase region with the maximum number of phases: $\alpha + \beta + \gamma$.

The cross rule is useful in checking the phases present in phase regions adjoining a point in two-dimensional diagrams. If one phase region is of known composition—say $l + \gamma$—then there are only two possible dispositions of the remaining phase regions (Fig. 234). The real utility of the concept of the cross rule and the law of adjoining phase regions is realised in the study of multicomponent systems**.

14.5. NON-REGULAR TWO-DIMENSIONAL SECTIONS

A section of a phase diagram is $(n+1) - \alpha$ dimensional, where α is the number of external factors assumed constant in the n-component phase diagram. We have ignored the pressure variable in discussing alloy phase equilibria and used the condensed Phase Rule. The phase diagrams considered throughout this book have been isobaric sections, the pressure being assumed to be atmospheric. Thus an isothermal section taken through an isobaric ternary system is $(n+1) - \alpha = (3+1) - 2 =$ two-dimensional. In the same way that the dimension of the section is less than that of the phase diagram, there is also a decrease in the dimension of the boundary between phase regions in the section compared to the phase diagram. In a space model of a ternary system (in reality an isobaric section) the boundaries between phase regions may be surfaces; in an isothermal or vertical section through the ternary the corresponding boundaries would be lines.

In regular sections the dimensions of the boundaries between phase regions are less than those in the corresponding phase diagram by a constant amount. In the case quoted above for a ternary system they would have one dimension less in the section compared to the phase diagram. Non-regular sections behave erratically and the dimensions of the phase region boundaries in such sections are reduced irregularly compared to those in the phase diagram. The phase boundaries of non-regular sections which are reduced in dimension less in the section than the dimensions of the phase diagram are reduced are called nodal plexi***. For example, comparing the three-dimensional ternary space model of Fig. 173a with the two-dimensional non-regular section of Fig. 178b the dimensions of the phase boundaries are uniformly one less in the section than those in the space model with the exception of the boundary represented by the invariant point E. This boundary exists as a point in both the space model and the non-regular section. The point E and the associated boundaries (Fig. 178b) is a nodal plexus. Note that the degenerate phase region $l + \alpha + \beta + \gamma$ is not shown in Fig. 178b.

---

  * L. S. PALATNIK AND A. I. LANDAU, *Zh. Fiz. Khim.*, **30** (1956) 2399.

  ** L. S. PALATNIK AND A. I. LANDAU, *Fazovye Ravnovesiya v Mnogokomponentnykh Sistemakh* (Phase Equilibria in Multicomponent Systems), Khar'kov State University, Khar'kov, 1961; translation edited by J. JOFFE, published by Holt, Rinehart and Winston, New York, 1964.

  *** L. S. PALATNIK AND A. I. LANDAU, *Zh. Neorg. Khim.*, **3** (1958) 637; *Russian J. Inorg. Chem.*, (Engl. Transl.) **3** (1958) 117.

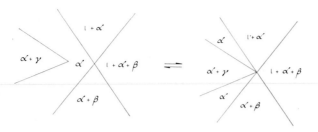

Fig. 235. Type 1 nodal plexus.

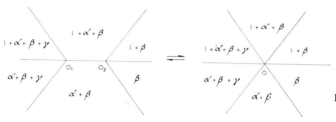

Fig. 236. Type 2 nodal plexus.

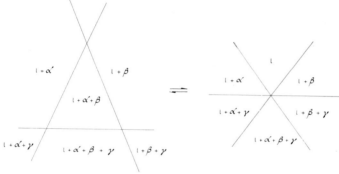

Fig. 237. Type 3 nodal plexus.

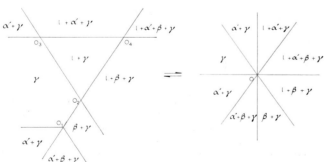

Fig. 238. Type 4 nodal plexus.

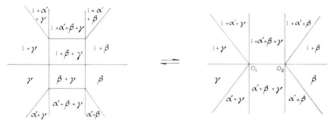

Fig. 239. Mixed type 2/3 nodal plexus.

Nodal plexi can be classified* into four types according to the manner of their formation:

Type 1. The nodal plexus is formed without degeneration of any geometrical element of the two-dimensional regular section to elements of a lower dimension (Fig. 235).

Type 2. A number of lines degenerate to a point but there is no degeneration of two-dimensional phase regions (Fig. 236). In the formation of a type 2 nodal plexus the line O_1O_2 in the regular section degenerates into point O of the nodal plexus associated with the non-regular section.

Type 3. A number of two-dimensional phase regions degenerate into a point (Fig. 237). In this case the phase region $l + \alpha + \beta$ disappears with the transition from a regular to a non-regular two-dimensional section.

Type 4. A number of two-dimensional phase regions degenerate to a line, (Fig. 238). In the formation of the nodal plexus the phase regions $l + \beta + \gamma$ and $\beta + \gamma$ have degenerated into the line O_1O_2.

Nodal plexi of mixed types may also be formed. A type 2/3 one is shown in Fig. 239. In the formation of the nodal plexus the two-dimensional $l + \gamma$ region degenerates to a point—triangle $O_2O_3O_4$ degenerates to point O—and the line O_1O_2 degenerates to the same point O. The former process corresponds to the formation of a type 3 nodal plexus and the latter to the formation of a type 2 nodal plexus.

Figures 235–238 illustrate four general methods for the formation of a nodal plexus by the transition from a regular section to a non-regular section of a ternary phase diagram. The subsequent transition from the non-regular section back to a regular section has been called by Palatnik and Landau the opening of the nodal plexus. Several examples of the formation and opening of nodal plexi have been included in the two-dimensional sections previously discussed. These are drawn together in Fig. 240. As will be seen, the type of nodal plexus is usually dependent on the direction from which the non-regular section is approached. In Fig. 240c the nodal plexus is type 3, irrespective of the direction of the transition. The rest of the examples show nodal plexi differing according to the direction of approach. In Fig. 240b there is a type 2 nodal plexus for the transition from left to right and a type 3 for the reverse transition.

Below each generalised example of the formation and opening of a nodal plexus in Fig. 240 is given an example from sections of ternary systems previously discussed. Figure 240d and e call for comment. In the formation of a type IV nodal plexus (Fig. 240d) the example quoted refers to vertical sections through the ternary system given in Fig. 185a–b. The three vertical sections (Fig. 241a–c) are constructed from the sections 1–2, 3–4 and 5–6. Figure 241d (*cf.* with Fig. 185b) gives the relative positions of the sections. It will be appreciated that section 3–4 (Fig. 241b) passes through both the invariant points b and c, and is therefore a non-regular section.

If Fig. 242 is combined with Fig. 225f and e, the sequence of sections provides an example of the formation and opening of a mixed-type nodal plexus. This sequence is identical with that given in Fig. 240e. The section shown in Fig. 242 is intermediate between Fig. 225f and g.

In general, the distribution of phases in non-regular sections does not obey the cross rule. Consider the non-regular section 11–12 through the invariant point c in Fig. 177. In the section

---

* L. S. PALATNIK AND A. I. LANDAU, *Fazovye Ravnovesiya v Mnogokomponentnykh Sistemakh* (Phase Equilibria in Multicomponent Systems), Khar'kov State University, Khar'kov, 1961, p. 270 *et seq.*

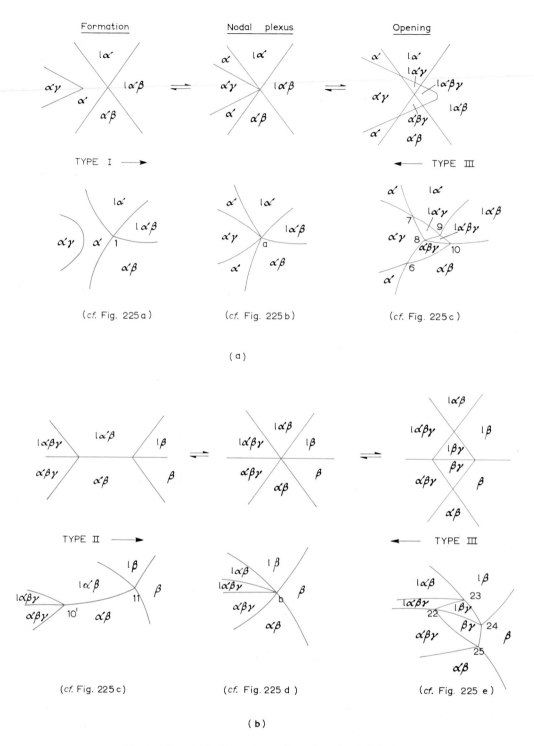

Fig. 240 (a and b). Formation and opening of nodal plexi.

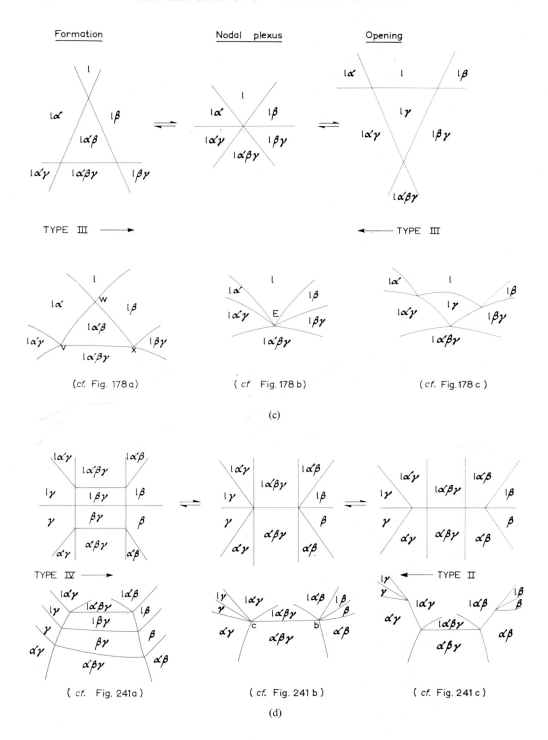

Fig. 240 (c and d).

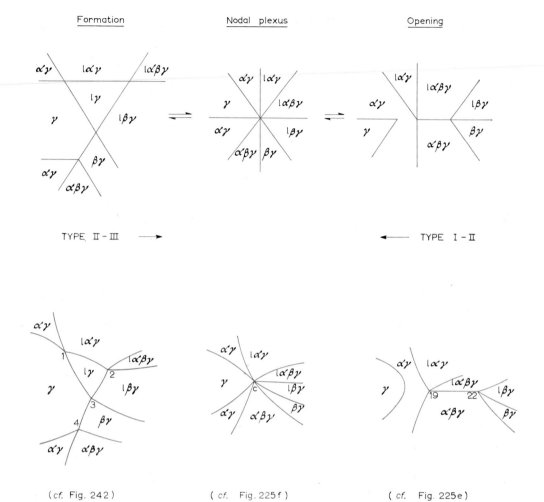

Fig. 240 (e).

(Fig. 178f) six lines meet at point c. Referring to the ternary space model in Fig. 173a, eight phase regions adjoin point c. These are:

(1) γ where c is a point on surface $T_c c_2 c_2^1 h c_3^1 c_3$

(2) $l+\gamma$ where c is a point on surface $T_c c_3 c c_2$

(3) $\alpha+\gamma$ where c is a point on surface $c_3 c h c_3^1$

(4) $\beta+\gamma$ where c is a point on surface $c_2 c h c_2^1$

(5) $l+\alpha+\gamma$ where c is a point on line $c_3 c$

(6) $l+\beta+\gamma$ where c is a point on line $c_2 c$

(7) $\alpha+\beta+\gamma$ where c is a point on line ch

(8) $l+\alpha+\beta+\gamma$ where c is a point representing one apex of the phase region.

Of these eight phase regions only six adjoin point c in the non-regular section (Fig. 178f). In the transition from the non-regular section to the regular sections which straddle it the other two phase regions will appear. These phase regions are the γ and the $l+\alpha+\beta+\gamma$ regions.

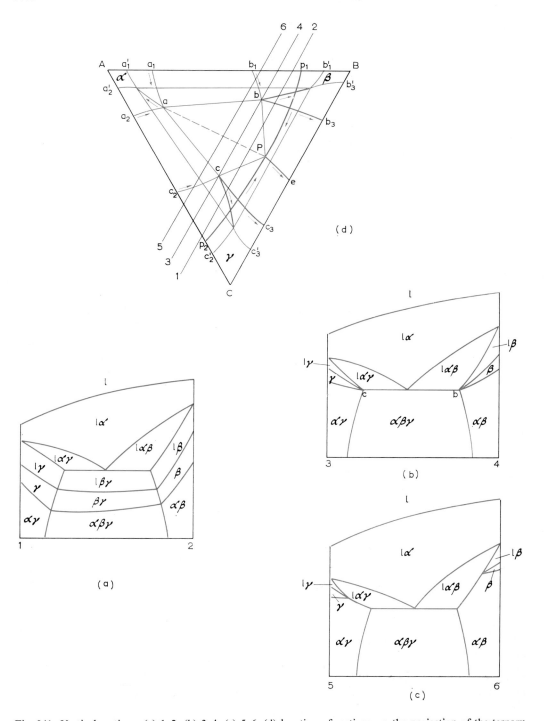

Fig. 241. Vertical sections, (a) 1–2, (b) 3–4, (c) 5–6, (d) location of sections on the projection of the ternary system (Fig. 185b).

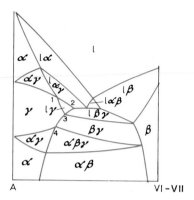

Fig. 242. Vertical section intermediate between Figs. 225f and g.

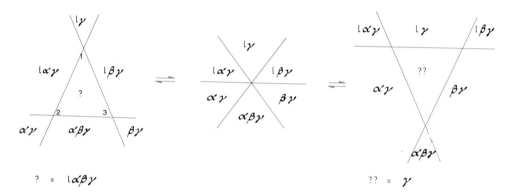

Fig. 243. Transition of the non-regular section (middle figure) into regular sections.

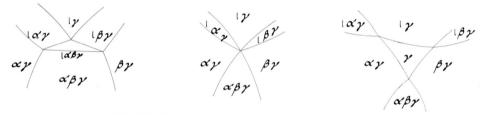

Fig. 244. Corresponding vertical sections to Fig. 243.

There are three methods by which a non-regular section of the type shown in Fig. 178f may change to a regular section. These methods are given in Figs. 240a, b and c. The nodal plexus of Fig. 240a is not applicable since two of the phase regions are identical, whereas in Fig. 178f all six phase regions have different phase compositions. Application of the law of adjoining phase regions to the regular sections formed from the central nodal plexus of Fig. 240b shows clearly that Fig. 178f cannot transform in this manner. This leaves Fig. 240c as the only possible mode of transition of the non-regular section 11–12. The transition is illustrated in Fig. 243. The phase region marked? is found to be the $l+\alpha+\beta+\gamma$ region by application of the law of adjoining phase regions to point 1. Checks can be made by applying the law to points 2 and 3. Phase region ?? can similarly be shown to be the γ phase region. The equivalent vertical sections have the form shown in Fig. 244. These should be compared with Fig. 178e, f and g.

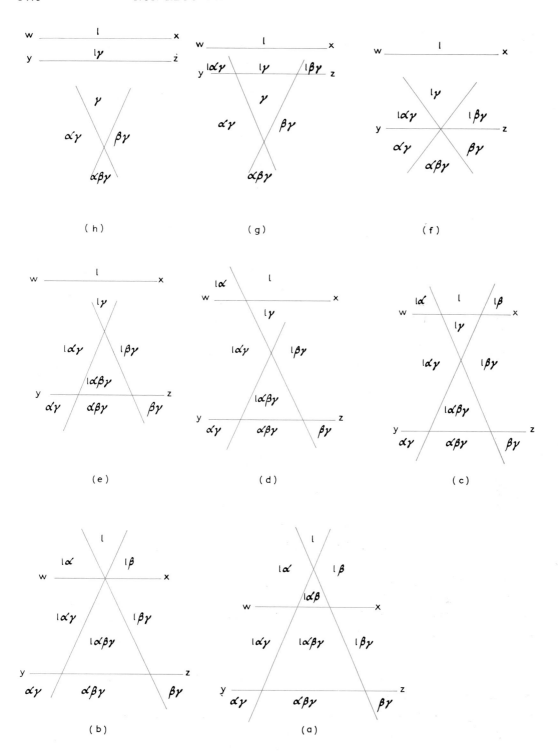

Fig. 245. Formation of the sequence of vertical sections (Fig. 178a–h) by the movement of lines *wx* and *yz*.

In the transition from the non-regular section to the regular section on the left, the $l\alpha\beta\gamma$ phase region appears as an additional phase region; in the transition to the regular section on the right the γ phase region appears as an additional phase region.

The importance of non-regular sections lies in the fact that they represent an intermediate step in the transition from one regular section to another regular section. If we started with the two non-regular sections 11–12 and 3–4 passing through the invariant points c and E of Fig. 177, we could construct the sequence of vertical sections given in Fig. 178. This can be shown by considering the complete vertical sections as, for example, in Fig. 245f. The sequence of sections (Fig. 245) can be viewed as being formed by the movement of lines wx and yz in a downwards direction.

14.6. CRITICAL POINTS

The rule of adjoining phase regions does not apply in the immediate neighbourhood of critical points in phase diagrams and their sections. Palatnik and Landau* have produced an empirical formula for the determination of the dimensions of a critical element: $R_1 = R - D_c \geqslant 0$, where R_1 is the dimension of the boundary between neighbouring phase regions, R is the dimension of the phase diagram or section, and D_c represents the number of phases that are combined into one phase as a result of the corresponding critical transformation.

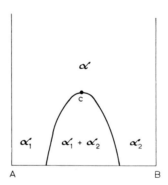

Fig. 246. Binary miscibility gap with critical point c.

The application of this formula can be illustrated by considering the case of a binary miscibility gap with a critical point c, (Fig. 246). According to the above formula, $R_1 = R - D_c = 2 - 2 = 0$. The critical element is zero-dimensional—point c. In this example $D_c = 2$ since the two phases α_1 and α_2 merge at the critical point into the α phase.

14.7. ASSOCIATION OF PHASE REGIONS IN MULTICOMPONENT SYSTEMS

The work of Palatnik and Landau as described in this chapter, and their further work on the construction of two-dimensional polythermal sections of phase diagrams** can be applied to the consideration of multicomponent systems. Reference is made to this in Chapter 15.

---

  * L. S. PALATNIK AND A. I. LANDAU, *Fazovye Ravnovesiya v Mnogokomponentnykh Sistemakh* (Phase Equilibria in Multicomponent Systems), Khar'kov State University, Khar'kov, 1961, pp. 227–233.
 ** L. S. PALATNIK AND A. I. LANDAU, *ibid.*, pp. 321–402; see also A. PRINCE, *Met. Rev.*, **8** (1963) 213.

Quaternary Phase Diagrams

15.1. REPRESENTATION OF QUATERNARY SYSTEMS

A quaternary system contains four components: A, B, C and D. Assuming isobaric conditions, the graphical representation of the constitution of a quaternary system requires the representation of four variables: X_A, X_B, X_C and T. The temperature–concentration diagram of a quaternary system is therefore represented by a four-dimensional figure. In view of the difficulty of four-dimensional geometry it is necessary to impose a further restriction on the system so as to allow a three-dimensional representation.

The most common figure used is the equilateral tetrahedron (Fig. 247) in which the apices represent the pure components A, B, C and D, the edges represent the six binary systems AB, AC, AD, BC, BD and CD, the triangular faces represent the four ternary systems ABC, ABD, ACD and BCD, and finally the interior of the tetrahedron represents the quaternary system ABCD. It is a property of an equilateral tetrahedron that lines drawn from an internal point parallel to four different edges of the tetrahedron so as to intersect each face of the tetrahedron,

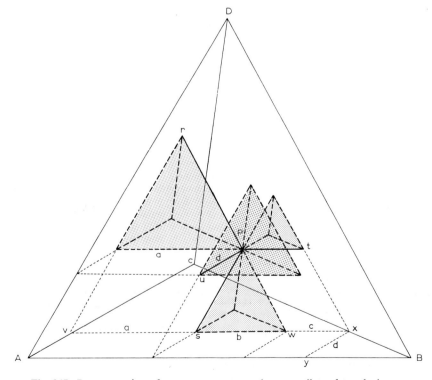

Fig. 247. Representation of a quaternary system by an equilateral tetrahedron.

have a total length equal to that of one edge of the tetrahedron. Referring to Fig. 247, a point P is taken to represent the composition of a quaternary alloy. From P six lines are drawn, parallel to the six sides of the tetrahedron, to meet the faces of the tetrahedron. In this way four small equilateral tetrahedra are formed with edge lengths of a, b, c and d. If the side AB has a length of 100 then

$$a+b+c+d = 100.$$

Since $vs = a$; $sw = b$ and $wx = c$

$$vs+sw+wx = vx = Ay.$$

$$\text{As } yB = xy = d,$$

then

$$AB = Ay+yB=a + b+c+d = 100.$$

The composition of alloy P is represented by the lengths a, b, c and d:

$$\%A = Pt = c,$$
$$\%B = Pr = a,$$
$$\%C = Pu = d,$$
$$\%D = Ps = b.$$

To plot the position of an alloy of known composition the $\%$ A is measured along any edge from the face opposite A, *i.e.* along either DA, CA, or BA from face BCD. In Fig. 248

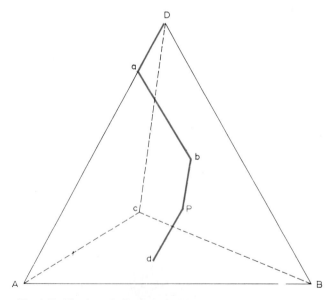

Fig. 248. Plotting of alloy compositions in quaternary systems.

we have chosen to plot $\%$ A $(= Da)$ along edge DA. From point a the $\%$ B $(= ab)$ is measured on the face ABD parallel to DB. From point b the $\%$ C $(= bP)$ is measured into the tetrahedron parallel to DC. The line from P to the basal triangle ABC is drawn parallel to the face ACD and represents $\%$ D $(= Pd)$.

Other useful geometrical properties of an equilateral tetrahedron are:

(1) Points on any line drawn from one apex to the opposite face represent alloys in which

there is a constant ratio of the components comprising the face. Alloys on line AE (Fig. 249a) have a variable A content but a fixed ratio of B : C : D.

(2) Alloys with a constant % A are situated on a plane parallel to the face of the tetrahedron which is opposite A. Alloys on plane FGH (Fig. 249b) contain a constant % A.

(3) All points on a plane including one edge of the tetrahedron correspond to alloys with a constant ratio of the other components. Alloys on plane ABJ (Fig. 249c) have a constant ratio of C : D.

(4) Planes, such as AJK in Fig. 249d, which originate at an apex and are parallel to an edge

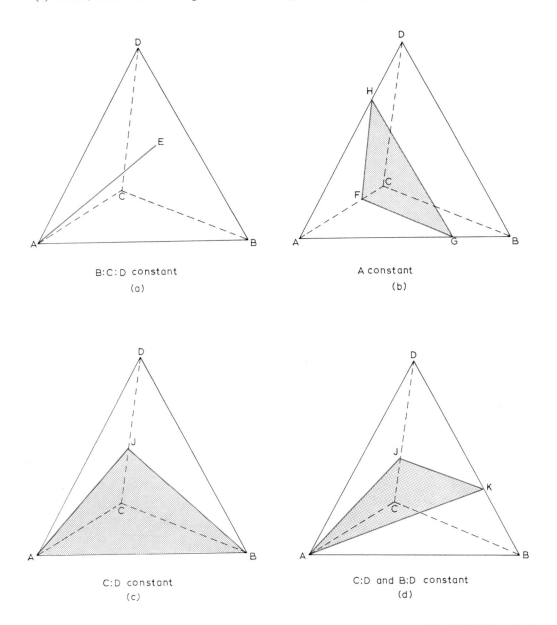

Fig. 249. Geometrical properties of an equilateral tetrahedron. (a) B : C : D constant on line AE; (b) A constant on plane FGH; (c) C : D constant on plane ABJ; (d) C : D and B : D constant on plane AJK.

on the opposite face of the tetrahedron contain alloys for which the ratios C : D and B : D are constant.

The three-dimensional equilateral tetrahedron can be used to study quaternary equilibria in the following ways.

Isobaric–isothermal sections

If both P and T are fixed, a quaternary system can be represented by the three concentration variables X_A, X_B and X_C (since $X_A + X_B + X_C + X_D = 100$). It is necessary to produce a series of three-dimensional tetrahedra to indicate the equilibria at a series of temperatures. Such isobaric–isothermal sections are equivalent to (isobaric) isothermal sections of ternary systems (*cf.* Fig. 176a–h).

Polythermal sections

In considering ternary systems use was made of a projection of the ternary space model on to the concentration triangle ABC to give a polythermal two-dimensional representation of the ternary system. Arrows on the monovariant curves in the projection indicated the direction in which these curves sloped with fall in temperature (*cf.* Fig. 173b). Similarly, quaternary systems can be represented by a polythermal projection which is three-dimensional.

Temperature–concentration sections are also useful. They can be either three-dimensional or two-dimensional. The latter are most frequently used. Referring to Fig. 249b, a three-dimensional temperature–concentration section can be constructed by fixing the %A and erecting temperature axes perpendicular to the plane FGH. Such a three-dimensional space model has a superficial resemblance to a ternary space model. It is really equivalent to a two-dimensional vertical section through a ternary system. As with such two-dimensional sections, tie lines do not generally lie wholly within the quaternary three-dimensional section but extend through a series of such three-dimensional sections.

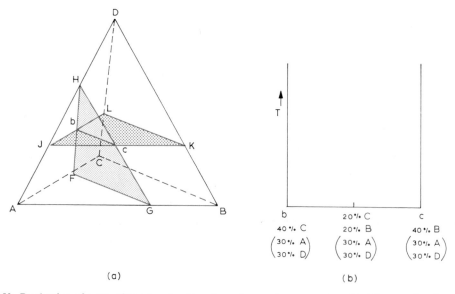

(a)   (b)

Fig. 250. Production of a two-dimensional section through a quaternary system. (a) Alloys on line *bc* contain constant content of A and D; (b) erection of temperature axis on line *bc* to give the two-dimensional section.

If the concentration of two components is fixed (say X_A and X_D), we are left with two variables —T and X_B. In Fig. 250a the concentration of components A and D has been fixed at 30% each. All alloys on the plane FGH contain 30 %A; all alloys on plane JKL contain 30 %D. Only alloys along the line of intersection, bc, of planes FGH and JKL contain 30 %A and 30 %D. Point b on the face ADC is a ternary alloy with 30 %A, 30 %D and 40 %C. Point c on the ternary face ADB is a ternary alloy with 30 %A, 30 %D and 40 %B. Erection of a temperature axis on line bc produces a two-dimensional temperature–concentration section of the quaternary (Fig. 250b).

The use of the above three-dimensional and two-dimensional methods for representing quaternary equilibria will be illustrated for simple quaternary systems. The systems will be classified according to whether the equilibrium is two-, three-, four- or five-phase.

15.2. TWO-PHASE EQUILIBRIUM

The most important two-phase equilibrium is that between melt and α solid solution: $l \rightleftharpoons \alpha$. As in the binary and ternary systems (Figs. 23f and 111), there will be a one-phase liquid region, a one-phase α region and, between them a two-phase $l+\alpha$ region. To clarify the comparison of ternary and quaternary systems two isobaric–isothermal sections are given in Fig. 251. Figure 251a is a ternary section and Fig. 251b a quaternary section. In the ternary section

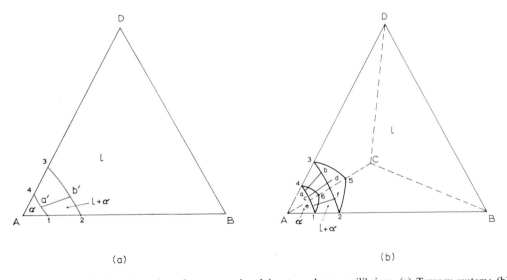

(a)   (b)

Fig. 251. Isobaric–isothermal sections for systems involving two-phase equilibrium. (a) Ternary system; (b) quaternary system.

a one-phase region, such as $B\,2\,3\,D\,B$ for the melt, is represented by an area. It makes contact with a two-phase region over a boundary line such as 2 3. A two-phase region, $l+\alpha$, is an area enclosed by two conjugate boundary lines (1 4 and 2 3). Tie lines such as $a'b'$, connect the conjugate phases on the boundary lines. In the quaternary isobaric–isothermal section a one-phase region, such as $B\,C\,D\,3\,5\,2\,B$, is represented by a space. It makes contact with a two-phase region over a boundary surface such as 3 5 2. A two-phase region, $l+\alpha$, is a space enclosed by two conjugate boundary surfaces (3 5 2 and 4 6 1). Tie lines connect all points of surface 3 5 2

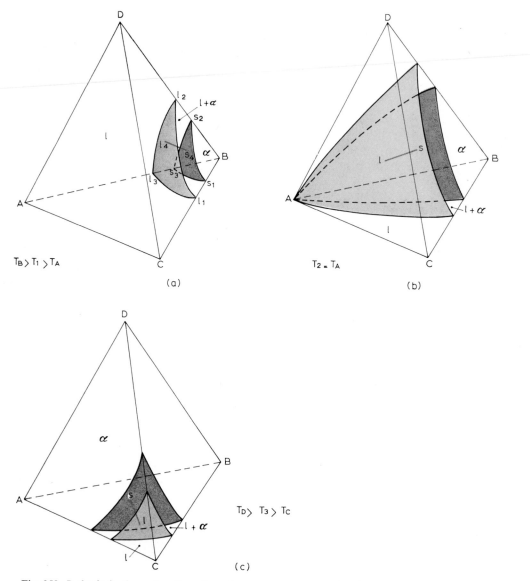

Fig. 252. Isobaric–isothermal sections through a quaternary system involving two-phase equilibrium.

to corresponding points on surface 4 6 1. The tie lines may run in any direction, *ab* or *cd* or *ef*, in the tetrahedral isotherm provided they do not intersect one another[*].

It will be noted that all geometrical elements increase their dimension by one in passing from a ternary to a corresponding quaternary system. Thus the lines and areas in Fig. 251a are converted into surfaces and state spaces in Fig. 251b.

The simplest quaternary system exhibiting two-phase equilibrium is the analogue of the ternary isomorphous system given in Fig. 111. All the binary systems have complete series of liquid and solid solutions and all four ternary systems are similar to Fig. 111. Three isobaric–isothermal sections are sufficient to indicate the nature of the quaternary system. It is assumed

---

[*] A. J. C. WILSON, *Proc. Cambridge Phil. Soc.*, **37** (1941) 95.

that the melting points of the components fall in the sequence B, A, D and C. Figure 252a is the isobaric–isothermal section for a temperature between T_B and T_A. On the ternary faces ABC, ABD, and BCD of the tetrahedron appear the liquidus and solidus lines formed by sectioning these ternary systems at the temperature of the quaternary section. Sectioning of the ternary system BCD leads to the liquidus curve l_1l_2 and the solidus curve s_1s_2. The three liquidus curves l_1l_2, l_2l_3 and l_3l_1 form a liquidus surface within the tetrahedron. The shape of this surface is similar to that of a jib sail with a gentle wind blowing from the direction of apex B. Similarly, for the solidus surface $s_1s_2s_3$. The α phase region is enclosed in the space between the solidus surface $s_1s_2s_3$ and apex B, the $l+\alpha$ phase region is enclosed between the liquidus and solidus surfaces, and the liquid phase region lies outside the liquidus surface $l_1l_2l_3$. A typical tie line is shown as line l_4s_4 where l_4 lies on the liquidus surface and s_4 on the solidus surface.

With fall in temperature the liquidus and solidus surfaces move outwards from the B corner and sweep towards the A corner. At the melting point of A (Fig. 252b) both the liquidus and solidus surfaces have stretched to apex A. The sections through ternary systems ABC and ABD have a form identical with that shown in Fig. 116a. Finally, in Fig. 252c the liquidus and solidus

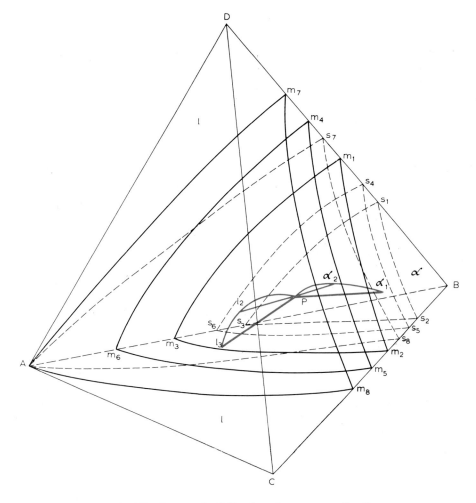

Fig. 253. Course of solidification of quaternary alloy P.

surfaces are bearing down rapidly on the C corner. As noted already, the quaternary tie lines in any one isobaric–isothermal section have a variety of orientations, but in going from one isothermal section to another with decreasing temperature the tie lines all change their orientation in such a manner that the quaternary melt is richer in the lower-melting components than the quaternary solid solution it is in equilibrium with (Konovalov's Rule). It should also be noted that the usual lever rule is applicable to tie lines in quaternary systems.

The solidification of any alloy in the quaternary system is analogous in many respects to the solidification of alloys in ternary isomorphous systems. In such systems projection on to the basal triangle of the tie lines connecting the conjugate liquid and α phases for an alloy X leads to the characteristic cusp in the curve representing the changes in composition of the liquid and α phases as solidification proceeds (cf. Fig. 121). In the corresponding quaternary system the solidification of an alloy P is represented by a projection of the composition changes of the conjugate liquid and α phases as shown in Fig. 253. Alloy P is assumed to begin solidifying at temperature T_1. The liquidus surface at temperature T_1 is surface $m_1m_2m_3$. Point P representing the alloy composition under consideration lies on surface $m_1m_2m_3$. The first solid to separate from liquid of composition P is of composition α_1 on surface $s_1s_2s_3$. The tie line $P\alpha_1$ is one of the many tie lines which connect conjugate phases on the liquidus surface $m_1m_2m_3$ and the solidus surface $s_1s_2s_3$. At a lower temperature T_2 the liquid composition l_2 lies on the liquidus surface $m_4m_5m_6$ and the conjugate solid α_2 on the solidus surface $s_4s_5s_6$. The tie line $l_2\alpha_2$ passes through the alloy composition P, but this tie line is not parallel to the first tie line $P\alpha_1$ in contrast to the series of tie lines in the ternary system (Fig. 121). It will have been noted that the liquidus and solidus surfaces $m_1m_2m_3$ and $s_1s_2s_3$ form the isobaric–isothermal section at a temperature T_1. Similarly, the surfaces $m_4m_5m_6$ and $s_4s_5s_6$ form the isobaric–isothermal section at T_2.

When the last drop of liquid is consumed at a temperature T_3 the liquid has a composition l_3 on the liquidus surface m_7m_8A and the conjugate solid is of composition P. Point P lies on the solidus surface s_7s_8A, i.e. point P lies on both the liquidus surface $m_1m_2m_3$ and the solidus surface s_7s_8A. In the solidification of alloy P the liquid traces curve Pl_2l_3 whilst the α phase follows the curve $\alpha_1\alpha_2P$. All quaternary alloys in this system solidify in a similar manner to alloy P. Any departures from equilibrium conditions during solidification results in the formation of a cored structure, as was noted for the binary and ternary alloys.

15.3. THREE-PHASE EQUILIBRIUM

In (isobaric) isothermal sections of ternary systems a three-phase equilibrium is represented by a tie triangle, the apices of which are the compositions of the co-existing phases. In Fig. 254a the tie triangle is marked $\alpha_1\beta_1l_1$, the α phase having composition α_1 and being in equilibrium with β of composition β_1 and liquid of composition l_1. In the corresponding isobaric–isothermal section of a quaternary system a three-phase equilibrium is represented by a space formed by three conjugate curves. Each pair of conjugate phases forms a ruled surface. In Fig. 254b a quaternary three-phase region is shown as being formed from a series of tie triangles such as $\alpha_1\beta_1l_1$, $\alpha\beta l$ and $\alpha_2\beta_2l_2$. Points α_1, α and α_2 on curve $\alpha_1\alpha_2$ represent the composition of the α phase in equilibrium with β along curve $\beta_1\beta_2$ and the liquid phase along curve l_1l_2. The $l\alpha$ ruled surface is $l_1l_2\alpha_2\alpha_1l_1$; the $l\beta$ ruled surface is $l_1l_2\beta_2\beta_1l_1$ and the $\alpha\beta$ ruled surface is $\alpha_1\alpha_2\beta_2\beta_1\alpha_1$. The tie triangles in the quaternary three-phase region do not lie parallel to each other since we are dealing with the series of conjugate phases which can be in equilibrium with each other

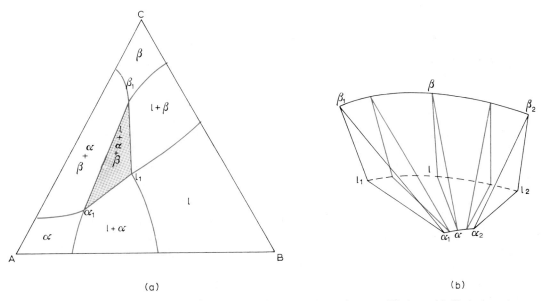

Fig. 254. Isobaric–isothermal sections for systems involving three-phase equilibrium. (a) Ternary system; (b) quaternary system.

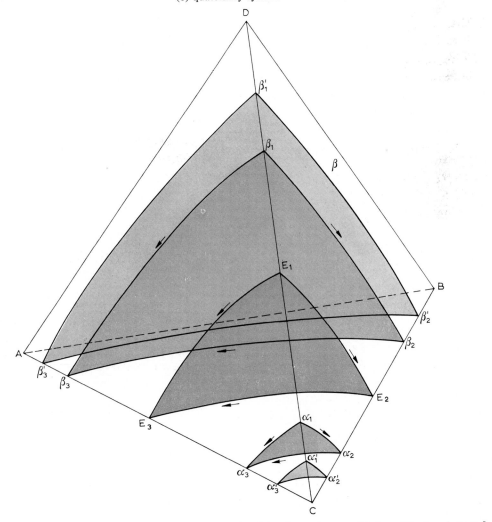

Fig. 255. Polythermal projection of a quaternary system involving three-phase equilibrium of the type $l \rightleftharpoons \alpha + \beta$

at a fixed temperature. This is in contrast to the superficially similar three-phase region in a ternary (isobaric) space model where the tie triangles remain parallel to each other as the temperature varies (*cf.* Fig. 134).

The simplest type of quaternary three-phase equilibrium is that involving the separation of two solids from a liquid phase: $l \rightleftharpoons \alpha + \beta$. In the system ABCD it is assumed that component C forms a binary eutectic with components A, B and D and that components A, B and D form continuous series of binary solid solutions with each other. Referring to the polythermal projection of the quaternary system (Fig. 255), face *ACD* of the tetrahedron is a polythermal projection of the ternary system ACD. There is a continuous transition from the binary eutectic CD to the binary eutectic AC through the ternary system ACD. The liquid follows the monovariant curve $E_1 E_3$ and the conjugate α and β phases trace curves $\alpha_1 \alpha_3$ and $\beta_1 \beta_3$ (*cf.* Fig. 164b). Faces *ABC* and *BCD* are similar to face *ACD*. No reactions are shown on the fourth face of the tetrahedron, face *ABD*, since this ternary system is isomorphous.

In the quaternary system there appear three surfaces representing the liquid, α and β phases ($E_1 E_2 E_3$, $\alpha_1 \alpha_2 \alpha_3$, and $\beta_1 \beta_2 \beta_3$). The change in solubility of the α and β phases with fall in temperature to room temperature is indicated by the additional α and β surfaces, $\alpha_1^1 \alpha_2^1 \alpha_3^1$ and $\beta_1^1 \beta_2^1 \beta_3^1$. Temperatures are assumed to decrease in this system in the sequence $T_D > T_C > E_1 > > T_B > T_A > E_2 > E_3$.

Isobaric–isothermal sections

If an isobaric–isothermal section is taken through the system at a temperature just below T_D the section will appear similar to Fig. 252a, with the exception that the one-phase region for existence of solid will be centred on apex D instead of apex B. Similarly, at a temperature just below T_C there will appear a solid phase region centred on the C apex. As the temperature falls further the liquidus surfaces growing outwards from apices D and C will approach each

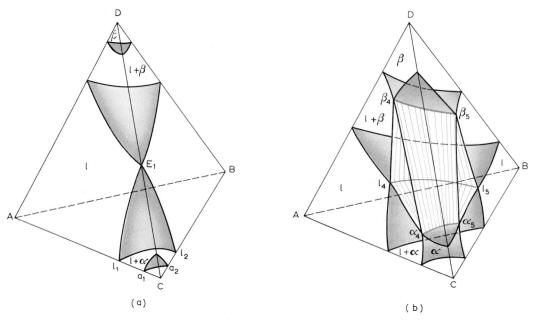

(a)   (b)

Fig. 256.

other. At temperature E_1 they will meet. They meet at the CD binary eutectic point E_1 (Fig. 256a). In constructing this section the binary eutectic point E_1 was first fixed by reference to the binary system CD. The ternary equilibria at the temperature of the binary eutectic E_1 was then drawn on faces BCD and ACD for the respective ternary systems. The ternary equilibria on faces ABD and ABC follow automatically. For example, binary tie lines l_1a_1 and l_2a_2 are fixed from the construction of faces BCD and ACD. Points l_1 and l_2, a_1 and a_2 are joined together on the face ABC to give the ternary ABC equilibria at temperature E_1. That this is the correct

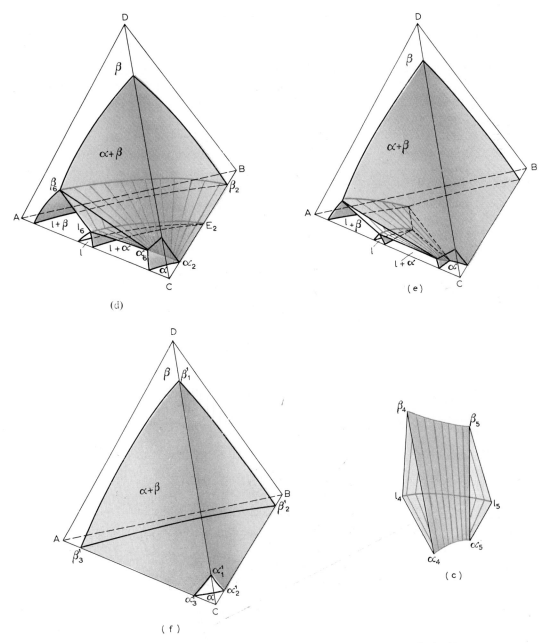

Fig. 256. Isobaric–isothermal sections through the quaternary system of Fig. 255.

equilibrium can be checked by recalling that $T_C > E_1 > T_B > T_A > E_2 > E_3$; only an area surrounding the C corner will be solid. Most of the ABC triangle will be liquid and the narrow strip of the $l+\alpha$ phase region will separate the liquid from the α region.

Below the temperature E_1 a three-phase triangle is observed in the ternary isothermal sections for systems ACD and BCD. The quaternary three-phase region is obtained by joining the two ternary tie triangles through the tetrahedron (Fig. 256b). The quaternary $l+\alpha+\beta$ phase region has the shape shown in Fig. 256c. The ternary tie triangles are labelled $l_4\alpha_4\beta_4$ and $l_5\alpha_5\beta_5$. With further fall in temperature the tie triangles on the ternary faces ACD and BCD move nearer to the AC and BC binary edges respectively. The corresponding quaternary three-phase region becomes broader and also approaches the ABC face of the tetrahedron. When the BC binary eutectic temperature is reached, E_2, the tie triangle on the BCD ternary face has degenerated into the tie line $\alpha_2E_2\beta_2$. The tie triangle on the ACD ternary face is now close to the AC edge. The quaternary three-phase region stretches from tie triangle $l_6\alpha_6\beta_6$ on the ACD ternary face to the degenerate tie line $\alpha_2E_2\beta_2$ on the binary BC edge of the tetrahedron (Fig. 256d).

At temperatures between E_2 and E_3 and $\alpha+\beta$ phase region stretches across the ternary face BCD. A tie triangle appears in the ternary ABC for the first time. This tie triangle is joined through the quaternary system to the tie triangle on the ternary face ACD (Fig. 256e).

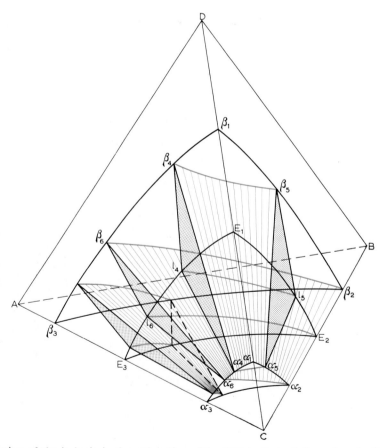

Fig. 257. Relation of the isobaric–isothermal sections (Fig. 256) to the polythermal projection of Fig. 255.

The liquid phase region has shrunk to a small tetrahedral space at this temperature. At the temperature of the AC binary eutectic, E_3, the liquid phase region degenerates into the point E_3 and the three-phase region degenerates into the binary eutectic tie line $\alpha_3 E_3 \beta_3$. At E_3 the last drop of liquid is consumed and all alloys in the quaternary system are completely solid at temperatures below E_3. The latter state of affairs is illustrated by Fig. 256f.

To assist in relating the three sections (Fig. 256b, d and e) containing the three-phase region to the polythermal projection given in Fig. 255, the three-phase regions from Fig. 256b, d and e have been superimposed on the polythermal projection in Fig. 257.

Equilibrium freezing of alloys

The solidification process for any quaternary alloy in this system can be studied with the aid of the series of isobaric–isothermal sections given in Fig. 256. This will give a picture of the sequence of phase regions which pass through the alloy point within the tetrahedron. An

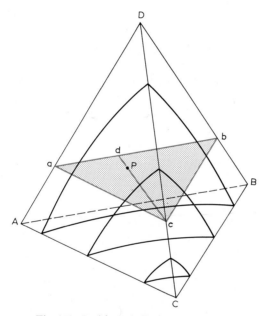

Fig. 258. Position of alloy P on plane *abc*.

alternative method uses a technique proposed by Schrader and Hannemann*. The essence of their method is the construction of a three-dimensional temperature–concentration section for a constant amount of one of the components. In Fig. 258 the section *abc* contains a constant %D. The section cuts across the ternary faces *ACD*, *ABD* and *BCD*. Vertical sections *a–b*, *a–c* and *c–b* of these ternary systems are easily constructed. They are shown in Fig. 259. Using these vertical sections a three-dimensional temperature–concentration diagram can be constructed for the quaternary system *abc* (Fig. 260a). A two-dimensional section *cd* is obtained by sectioning Fig. 260a, as illustrated in Fig. 260b for the section *c–d* through the quaternary system.

Consider the solidification of alloy P on plane *abc*, Figs. 258 and 260. The liquidus is reached

---

* A. Schrader and H. Hannemann, *Z. Metallk.*, **35** (1943) 185.

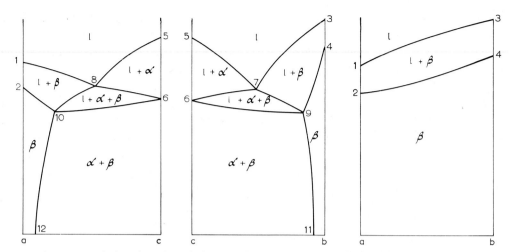

Fig. 259. Vertical sections through the ternary systems. a–c, system ACD; a–b, ABD; b–c, BCD.

at point s in Fig. 260b. This two-dimensional section tells us that the liquid is in equilibrium with β initially. After the primary solidification of β the liquid becomes saturated with α and both α and β separate together. When all the liquid has been consumed the alloy is two-phase $\alpha + \beta$. To gain an insight into the course of freezing of alloy P recourse must be had to a plot of the changes in composition of the liquid, α and β phases on the polythermal projection (Fig. 261). The liquid initially has a composition represented by point P. It begins to solidify by precipitating β solid solution of composition β_4. Note that the point β_4 does not lie on surface $\beta_1\beta_2\beta_3$ nor does it appear in the section given in Fig. 260b since it lies above the plane

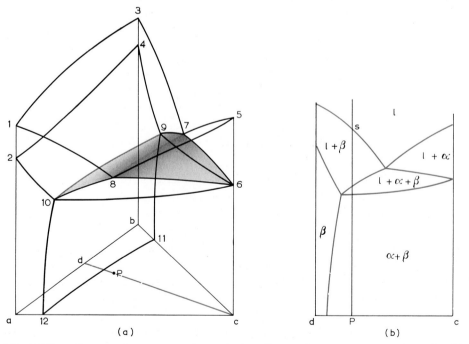

Fig. 260. (a) Three-dimensional temperature–concentration diagram for a quaternary system abc; (b) two-dimensional section through Fig. 260a.

abc. In fact the tie line $P\beta_4$ is the type of tie line shown as l_4s_4 in Fig. 252a. Point P lies on the liquidus surface and point β_4 on the solidus surface. With fall in temperature the compositions of the liquid and β phases change, but always in such a way that the tie line indicating the compositions of the conjugate liquid and β passes through point P. The liquid and β phases trace paths similar to those shown in Fig. 253. Eventually a temperature will be reached at which the liquid will meet the eutectic surface $E_1E_2E_3$. At this temperature liquid of composition l_5 is in equilibrium with β of composition β_5 and an infinitely small amount of α of composition α_5. The co-precipitation of α and β from the liquid has begun. Whereas the liquid of composition l_5 lies on the surface $E_1E_2E_3$, the conjugate α and β phases lie on surfaces $\alpha_1\alpha_2\alpha_3$ and β_1 $\beta_2\beta_3$. With the separation of both α and β simultaneously a three-phase equilibrium appears

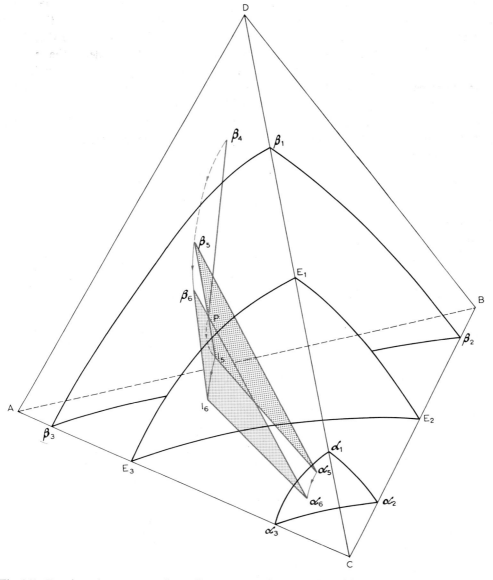

Fig. 261. Freezing of quaternary alloy P illustrated by reference to the polythermal projection of Fig. 255.

and this is represented by a tie triangle: $l_5\alpha_5\beta_5$. Tie line $l_5\beta_5$ of the tie triangle passes through point P. The three-phase equilibrium is maintained until the temperature has fallen to that represented by tie triangle $l_6\alpha_6\beta_6$. At this stage the tie line $\alpha_6\beta_6$ passes through point P and liquid is no longer taking part in the equilibrium. The alloy P is completely solid. Points l_5, l_6, α_5, α_6 and β_5, β_6 lie on surfaces $E_1E_2E_3$, $\alpha_1\alpha_2\alpha_3$ and $\beta_1\beta_2\beta_3$ respectively. As the last drop of liquid freezes the alloy P consists of the α phase of composition α_6 and the β phase of composition β_6. With cooling to room temperature the relative amounts of α and β vary since the α solubility surface $\alpha_1\alpha_2\alpha_3$ approaches the apex C and the β solubility surface approaches face ABD. The final positions of the α and β solubility surfaces are given in Fig. 255 as $\alpha_1^1\alpha_2^1\alpha_3^1$ and $\beta_1^1\beta_2^1\beta_3^1$ respectively.

In this quaternary system there are obviously two regions of primary crystallisation and one region of secondary crystallisation. Primary β precipitates from the liquid if the alloy in question has a composition within the volume between face ABD and surface $E_1E_2E_3$; primary α separates if the alloy has a composition within the volume between surface $E_1E_2E_3$ and apex C. Alloys within the space between surfaces $\alpha_1\alpha_2\alpha_3$ and $\beta_1\beta_2\beta_3$ undergo the secondary three-phase reaction involving simultaneous separation of α and β from the liquid. Alloys within the space between face ABD and $\beta_1^1\beta_2^1\beta_3^1$ undergo the two-phase reaction $l \rightleftharpoons \beta$ only; alloys between surfaces $\beta_1^1\beta_2^1\beta_3^1$ and $\beta_1\beta_2\beta_3$ solidify initially as β but, at lower temperatures, some α separates from the β. Finally, alloys within the tetrahedron $C\,\alpha_1^1\alpha_2^1\alpha_3^1$ solidify as α and alloys between surfaces $\alpha_1^1\alpha_2^1\alpha_3^1$ and $\alpha_1\alpha_2\alpha_3$ behave similarly but some β separates from the α at lower temperatures.

15.4. FOUR-PHASE EQUILIBRIUM

In (isobaric) isothermal sections of ternary systems a four-phase equilibrium is represented by a degenerate tie-tetrahedron. Four phases can only exist at one temperature in a ternary system. Hence four phases only occur in one of a series of isothermal sections (Fig. 262). These are the phases: liquid, α, β and γ, of composition E, a, b and c respectively, for the ternary eutectic reaction illustrated.

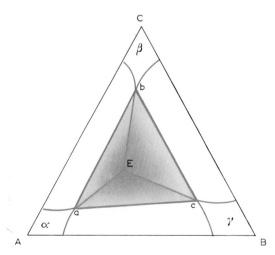

Fig. 262. Isobaric–isothermal section of a ternary system at the ternary eutectic temperature.

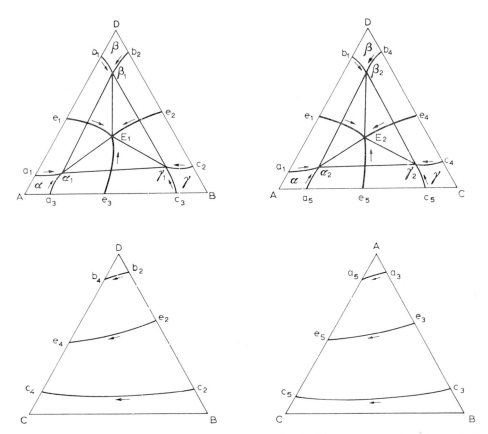

Fig. 263. Polythermal projections of the ternary systems involved in a quaternary four-phase eutectic equilibrium.

In the corresponding isobaric–isothermal section of a quaternary system a four-phase equilibrium is represented by a tie-tetrahedron. The apices of the tetrahedron represent the compositions of the four co-existing phases; any alloy whose composition lies within the tie tetrahedron will separate into the four phases. Three-phase regions adjoin the four-phase tie tetrahedron over the tetrahedral faces, two-phase regions along the edges and single phase regions at the apices. If the equilibrium involves liquid, α, β and γ phases, one tie tetrahedron, with these phases at the apices, will appear in each appropriate isobaric–isothermal section. To trace the course of the four-phase equilibrium we must consider a series of isobaric–isothermal sections.

Hillert's criterion can be applied to quaternary systems in a similar manner to its application to ternary three-phase equilibria (p. 154). In the case of a quaternary four-phase equilibrium application of the criterion* indicates that:

(1) If Δm_α, Δm_β and Δm_γ are positive and Δm_l is negative, the quaternary four-phase equilibrium is of the eutectic type: $l \rightleftharpoons \alpha + \beta + \gamma$.

(2) If one the expressions Δm_α, Δm_β and Δm_γ, is negative and Δm_l is negative, the quaternary reaction is a quasi-peritectic type: $l + \alpha \rightleftharpoons \beta + \gamma$ for Δm_α negative.

---

* R. CAYRON, *Etude Theorique des Diagrammes d'Equilibre dans les Systemes Quaternaires*, Institut de Metallurgie, Louvain, 1960.

(3) If two of the expressions Δm_α, Δm_β and Δm_γ, are negative and Δm_l is negative, the quaternary reaction is a peritectic type: $l + \alpha + \beta \rightleftharpoons \gamma$ for Δm_α and Δm_β negative.

The simplest case of quaternary four-phase equilibrium involves the eutectic reaction, $l \rightleftharpoons \alpha + \beta + \gamma$. The corresponding quaternary system is constructed from five binary eutectic systems—AB, AC, AD, CD and BD—and one binary solid solution system—BC. The ternary systems ABD and ACD are ternary eutectic types, and the ternary systems ABC and BCD show only ternary three-phase equilibria (Fig. 263). In the polythermal section (Fig. 264) temperatures are assumed to decrease in the following order:

$$T_D > T_A > T_B > e_1 > e_2 > T_C > e_3 > e_4 > e_5 > E_1 > E_2,$$

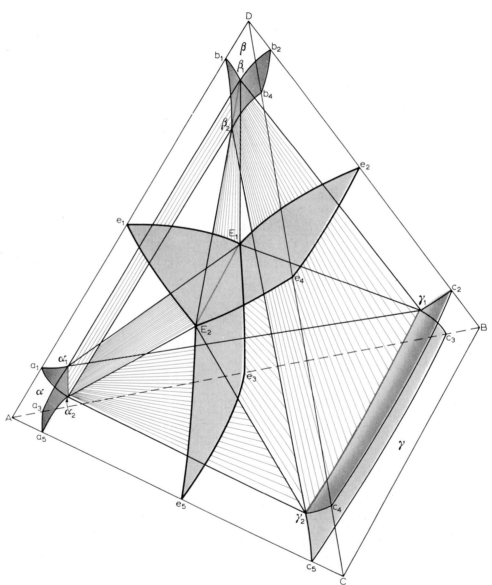

Fig. 264. Polythermal projection of a quaternary system involving four-phase equilibrium of the type $l \rightleftharpoons \alpha + \beta + \gamma$.

where $e_1, \ldots, e_5$ are the binary eutectic temperatures for systems AD, BD, AB, CD and AC, and E_1 and E_2 are the ternary eutectic temperatures for systems ABD and ACD. Changes in solid solubility are ignored for the sake of greater clarity.

In essence, the polythermal section shows that the ABD ternary eutectic reaction, $E_1 \rightleftharpoons \alpha_1 + \beta_1 + \gamma_1$, is converted into the monovariant quaternary eutectic reaction, $l \rightleftharpoons \alpha + \beta + \gamma$, in the quaternary system. The quaternary reaction ends at the ACD face, where it degenerates into the ACD ternary eutectic reaction, $E_2 \rightleftharpoons \alpha_2 + \beta_2 + \gamma_2$. Surfaces $e_1 E_1 E_2 e_1$, $e_2 E_1 E_2 e_4 e_2$ and $E_1 E_2 e_5 e_3 E_1$ divide the system into three regions of primary crystallisation. The space $e_1 E_1 e_2 e_4 E_2 e_1 D$ is the region for primary crystallisation of β, space $e_1 E_1 e_3 e_5 E_2 e_1 A$ for primary α, and space $e_2 e_4 E_2 E_1 e_3 e_5 CB$ for primary γ. Secondary crystallisation involves the equilibrium of three phases, e.g. l, α and β, as a series of tie triangles. The three phases are represented by points on conjugate curves. In Fig. 264 the $l \rightleftharpoons \alpha + \beta$ equilibrium is represented by the space between the two ternary tie triangles $a_1 \beta_1 E_1$ and $\alpha_2 \beta_2 E_2$. The conjugate curves are $E_1 E_2$ for the liquid phase, $\alpha_1 \alpha_2$ for the α phase and $\beta_1 \beta_2$ for the β phase. Similarly, the $l \rightleftharpoons \alpha + \gamma$ region of secondary crystallisation is represented by the space between tie triangles $\alpha_1 \gamma_1 E_1$ and $\alpha_2 \gamma_2 E_2$, and the $l \rightleftharpoons \beta + \gamma$ region of secondary crystallisation by the space between tie triangles $\beta_1 \gamma_1 E_1$ and $\beta_2 \gamma_2 E_2$.

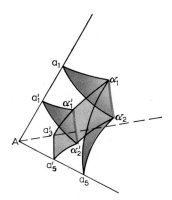

Fig. 265. Change in solid solubility for the α phase.

Tertiary crystallisation according to the equilibrium $l \rightleftharpoons \alpha + \beta + \gamma$ is represented by a series of tie tetrahedra the apices of which lie on the curves $E_1 E_2$, $\alpha_1 \alpha_2$, $\beta_1 \beta_2$ and $\gamma_1 \gamma_2$. After solidification is completed we are left with the equilibrium between α, β and γ, represented by points on curves $\alpha_1 \alpha_2$, $\beta_1 \beta_2$ and $\gamma_1 \gamma_2$. Changes in solid solubility of α, β and γ with fall in temperature to room-temperature are not indicated in Fig. 264. The construction for the α phase is given in Fig. 265. The room-temperature section has been added to give the change in solubility of α with fall in temperature. This change in solubility is represented by the surface $\alpha_1 \alpha_2 \alpha_2^1 \alpha_1^1 \alpha_1$. Similar surface can be constructed for the β and γ phases.

Isobaric–isothermal sections

The isobaric–isothermal section at a temperature e_1 (Fig. 266a) is similar to Fig. 256a. Below e_1 a three-phase $(l + \alpha + \beta)$ region appears in the ternary systems ACD and ABD. They are connected in the quaternary system by curves $\alpha_3 \alpha_4$, $\beta_3 \beta_4$, and $l_3 l_4$ at a temperature assumed to be that of the binary BD eutectic, e_2 (Fig. 266b). The situation is similar to that previously depicted in Fig. 256b. At temperatures below e_2 a three-phase $(l + \beta + \gamma)$ region will appear on the ternary faces ABD and BCD. At temperature e_3 there will therefore be two quaternary three-phase regions representing equilibrium between l, α and β on the one hand and l, β and γ on the other, as shown in Fig. 266c.

When the temperature falls below e_3 a $(l+\alpha+\gamma)$ phase region appears on the ternary faces ABD and ABC. They are connected in the quaternary system by curves $\beta_5\beta_6$, $\gamma_5\gamma_6$ and l_5l_6 if a temperature e_4 is considered (Fig. 266d). At this temperature the liquid phase is restricted to a space which funnels from a triangular region within the ABD face to the rectangular region on the ACD face with a small triangular region on the ABC face. There are now three quaternary three-phase regions—$l+\alpha+\beta$ stretching from the ACD to the ABD face, $l+\beta+\gamma$

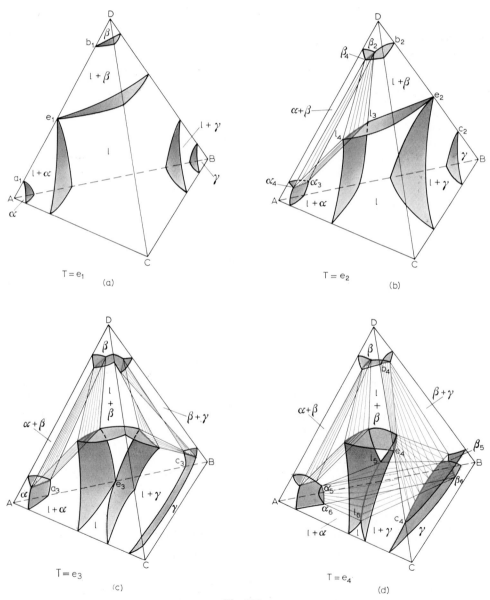

Fig. 266

stretching from the degenerate tie line $b_4e_4c_4$ of the CD binary to the ABD face, and $l+\alpha+\gamma$ stretching from the ABC face to the ABD face.

As the temperature falls to e_5 the $l+\alpha+\gamma$ phase region on the ABC face will move towards the AC edge. At e_5 it will degenerate into the tie line $a_5e_5c_5$ on edge AC. Below e_5 the $l+\alpha+\gamma$ phase region will make its first appearance on face ACD. If the temperature falls below the ternary eutectic temperature E_1 on face ABD (Fig. 266e), an $\alpha+\beta+\gamma$ region will appear on face ABD. This phase region will extend into the quaternary from the ternary tie triangle $\alpha_7\beta_7\gamma_7$ to the $\alpha\beta\gamma$ surface $\alpha_8\beta_8\gamma_8$ such that α, β and γ are in equilibrium along curves $\alpha_7\alpha_8$, $\beta_7\beta_8$ and $\gamma_7\gamma_8$. Surface $\alpha_8\beta_8\gamma_8$ is one surface of the tie tetrahedron which represents equilibrium between l, α, β and γ. The apices of the $l+\alpha+\beta+\gamma$ tie tetrahedron are points l_8, α_8, β_8 and γ_8. As noted previously, points l_8, α_8, β_8 and γ_8 lie on curves E_1E_2, $\alpha_1\alpha_2$, $\beta_1\beta_2$ and $\gamma_1\gamma_2$ (Fig. 264). The $l+\alpha+\beta+\gamma$ tetrahedron originates on the ternary face ABD at a temperature E_1 and moves into the quaternary system towards face ACD as the temperature falls from E_1 to E_2. The liquid region (Fig. 266e) is now a curved tetrahedron based on the ternary face ACD.

The dimensions of the boundary between neighbouring phase regions in quaternary isobaric–isothermal sections can be obtained by applying the rule of adjoining phase regions*. It will then be found that the boundaries have dimensions identical with those stated on p. 235 for a three-dimensional model of a ternary system. Adjoining each m-dimensional geometrical element in a regular section of a phase diagram there are 2^{R-m} phase regions, where R is again the dimension of the phase diagram or regular section. For an isobaric–isothermal regular section of a quaternary phase diagram eight phase regions will adjoin a point ($m = 0$), four adjoin a line ($m = 1$) and two have a common surface ($m = 2$). By this means a simple check can be made that the correct number of phase regions have been located in any regular section.

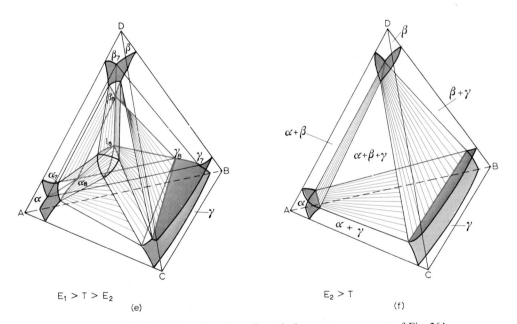

$E_1 > T > E_2$

(e)

$E_2 > T$

(f)

Fig. 266. Isobaric–isothermal sections through the quaternary system of Fig. 264.

---

* The geometrical rules which must be obeyed in three-dimensional quaternary phase diagrams were initially summarised by A. J. C. WILSON (*Proc. Cambridge Phil. Soc.*, **37** (1) (1941) 95).

The final section is taken at a temperature below E_2. All alloys in the system are solid and the $\alpha + \beta + \gamma$ phase region stretches from the ternary face ABD to the ternary face ACD (Fig. 266f).

Vertical sections at constant %D

In establishing the vertical section (Fig. 267c) it is necessary to construct the three ternary sections a–b, a–c and b–c (Fig. 267b). Their location relative to the quaternary section at

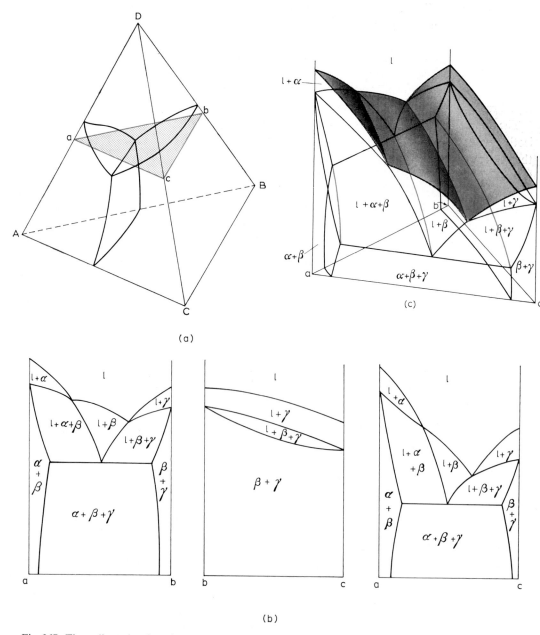

Fig. 267. Three-dimensional section at constant % D. (a) Location of alloys under consideration; (b) vertical sections of the ternary systems; (c) quaternary temperature–concentration section.

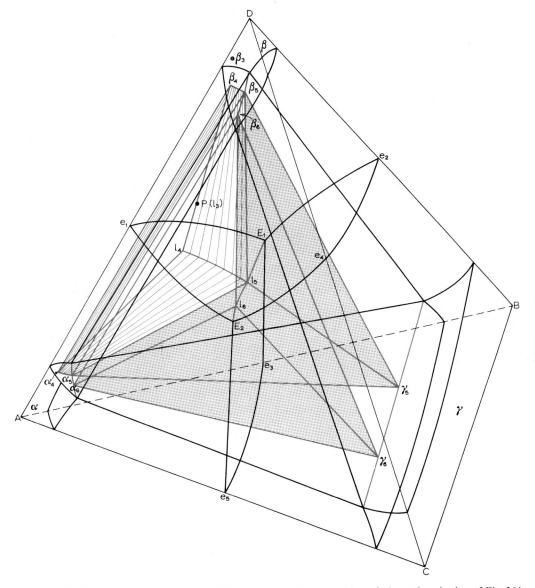

Fig. 268. Freezing of quaternary alloy *P* illustrated by reference to the polythermal projection of Fig. 264.

50 %D is illustrated in Fig. 267a. Two-dimensional vertical sections of this quaternary section can be taken as was done for Fig. 260. An alternative method for constructing two-dimensional sections of quaternary systems is given by Landau*.

Equilibrium freezing of alloys

Alloy *P* (Fig. 268) is in the primary β region. Solidification begins when the liquidus is reached at point *P*. Liquid of composition $l_3(= P)$ is in equilibrium with the initial precipitate of β of composition β_3. With further solidification the liquid composition traces a curved path until it reaches the binary eutectic-type surface $e_1E_1E_2e_1$ at point l_4. There is a corresponding change in composition of the β phase from β_3 to β_4. The latter composition lies on the β

* A. I. LANDAU, *Zh. Fiz. Khim.*, **36** (1962) 463; see also pp. 260–276 of the review by A. PRINCE (*Met. Rev.*, **8** (1963) 213).

surface $b_1\beta_1\beta_2$. When the liquid reaches the eutectic surface it becomes saturated with α as well as the β phase. The α has a composition α_4 and lies on the α surface $a_1\alpha_1\alpha_2$. At this temperature alloy P is composed of liquid l_4 in equilibrium with α and β of compositions α_4 and β_4 respectively. With fall in temperature the liquid, α and β compositions move over the conjugate surfaces $e_1E_1E_2e_1$, $a_1\alpha_1\alpha_2$ and $b_1\beta_1\beta_2$ tracing paths l_4l_5, $\alpha_4\alpha_5$ and $\beta_4\beta_5$. In this way the tie triangle $l_4\alpha_4\beta_4$ moves to $l_5\alpha_5\beta_5$. At this temperature the liquid is on the ternary eutectic-type curve E_1–E_2 and phases α and β are on the conjugate curves $\alpha_1\alpha_2$ and $\beta_1\beta_2$. The liquid is also conjugate with the γ curve $\gamma_1\gamma_2$ so it becomes saturated with γ of composition γ_5 in addition to α and β. The four-phase equilibrium between l, α, β and γ is now established. Point l_5 is the upper apex of the tetrahedron $l_5\alpha_5\beta_5\gamma_5$. As already noted when considering isobaric–isothermal sections, the tie tetrahedron moves towards the ternary eutectic point E_2 with further fall in temperature; the liquid moves along the curve l_5E_2, α along curve $\alpha_5\alpha_2$, β along $\beta_5\beta_2$ and γ along $\gamma_5\gamma_2$. Solidification is completed when the last drop of liquid of composition l_6 is consumed in depositing α, β and γ of compositions α_6, β_6 and γ_6. When this occurs the plane of the triangle $\alpha_6\beta_6\gamma_6$ includes the alloy point P within its interior. Further changes after solidification only involve a change in the relative amounts of α, β and γ with fall in temperature to room-temperature.

15.5. FIVE-PHASE EQUILIBRIUM

In an isobaric section of a quaternary system five phases can only exist in equilibrium at one temperature. Five-phase equilibrium is invariant. In a series of isobaric–isothermal sections a five-phase equilibrium is only found in the section taken at the temperature of the invariant reaction. With liquid, α, β, γ and δ phases present the invariant reaction can be one of four types:

(1) $l \rightleftharpoons \alpha+\beta+\gamma+\delta$ quaternary eutectic,
(2) $l+\alpha \rightleftharpoons \beta+\gamma+\delta$ 2-3 quaternary quasi-peritectic,
(3) $l+\alpha+\beta \rightleftharpoons \gamma+\delta$ 3-2 quaternary quasi-peritectic,
(4) $l+\alpha+\beta+\gamma \rightleftharpoons \delta$ quaternary peritectic.

In each case the compositions of the five phases and the reaction temperatures are fixed.

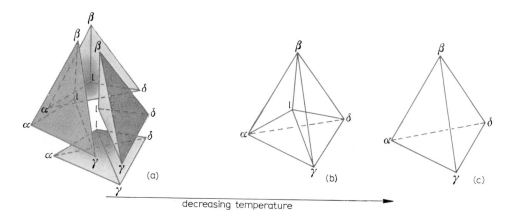

decreasing temperature

Fig. 269. Sequence of tie-tetrahedra on cooling through the quaternary eutectic temperature.

(1) *Quaternary eutectic*

The compositions of the five co-existing phases are denoted (Fig. 269b) by points l, α, β, γ and δ. The liquid composition lies within the tetrahedron formed by the four solid phases. In a series of isobaric–isothermal sections four tie tetrahedra — $l\alpha\beta\gamma$, $l\alpha\beta\delta$, $l\alpha\gamma\delta$ and $l\beta\gamma\delta$ — will exist at temperatures above the quaternary eutectic temperature (Fig. 269a). Below the eutectic temperature (Fig. 269c), only one tie tetrahedron exists — the $\alpha\beta\gamma\delta$ tetrahedron.

(2) *2-3 Quaternary quasi-peritectic*

A reaction of the type $l+\alpha \rightleftharpoons \beta+\gamma+\delta$ is represented by a tie hexahedron (Fig. 270b). It is constructed of six triangles, $l\beta\gamma$, $l\gamma\delta$, $l\beta\delta$, $\alpha\beta\gamma$, $\alpha\gamma\delta$ and $\alpha\beta\delta$. The points l and α lie on opposite

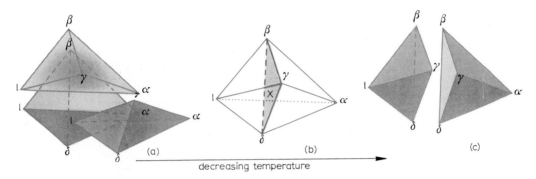

Fig. 270. Sequence of tie-tetrahedra on cooling through the quaternary 2-3 quasi-peritectic temperature.

sides of the triangle $\beta\gamma\delta$ and the line $l\alpha$ intersects the $\beta\gamma\delta$ surface at point X. The reaction $l+\alpha \rightleftharpoons \beta+\gamma+\delta$ can then be considered as proceeding in two steps:

$$l+\alpha \rightleftharpoons X$$
$$X \rightleftharpoons \beta+\gamma+\delta$$
$$\overline{l+\alpha \rightleftharpoons \beta+\gamma+\delta}$$

Above the invariant temperature there are three tie tetrahedra: $l\alpha\beta\gamma$, $l\alpha\beta\delta$ and $l\alpha\gamma\delta$ (Fig. 270a). Below it there are two tie tetrahedra: $\alpha\beta\gamma\delta$ and $l\beta\gamma\delta$ (Fig. 270c).

The term 2-3 quaternary quasi-peritectic is used to define the five-phase equilibrium in which two phases react to produce three new phases.

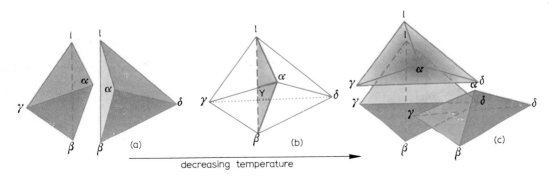

Fig. 271. Sequence of tie-tetrahedra on cooling through the quaternary 3-2 quasi-peritectic temperature.

(3) 3-2 Quaternary quasi-peritectic

The reaction $l + \alpha + \beta \rightleftharpoons \gamma + \delta$ is also represented by a tie hexahedron (Fig. 271b). In this case the tie triangle $l\alpha\beta$ intersects the line $\gamma\delta$ at point Y, and the invariant reaction can again be considered as proceeding in two steps:

$$l + \alpha + \beta \rightleftharpoons Y$$
$$\underline{Y \rightleftharpoons \gamma + \delta}$$
$$l + \alpha + \beta \rightleftharpoons \gamma + \delta.$$

Above the invariant temperature there are two tie tetrahedra: $l\alpha\beta\delta$ and $l\alpha\beta\gamma$ (Fig. 271a). Below it there are three tie tetrahedra: $l\alpha\gamma\delta$, $l\beta\gamma\delta$ and $\alpha\beta\gamma\delta$ (Fig. 271c).

(4) Quaternary peritectic

In a quaternary peritectic reaction, $l + \alpha + \beta + \gamma \rightleftharpoons \delta$ point δ lies within the tetrahedron $l\alpha\beta\gamma$ (Fig. 272b). Above the invariant temperature there is only one tie tetrahedron: $l\alpha\beta\gamma$ (Fig. 272a). Below it there are four tie tetrahedra: $l\alpha\beta\delta$, $l\alpha\gamma\delta$, $l\beta\gamma\delta$ and $\alpha\beta\gamma\delta$ (Fig. 272c).

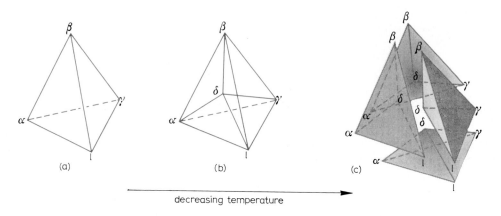

decreasing temperature

Fig. 272. Sequence of tie-tetrahedra on cooling through the quaternary peritectic temperature.

Only the quaternary eutectic five-phase equilibrium will be considered. The first quaternary system to be investigated was of this type*. The metals Bi, Cd, Pb and Sn form binary eutectic systems with each other. The ternary systems formed by these four metals are also simple eutectics. The ternary eutectic temperatures quoted by Parravano and Sirovich are 92 °C for the Bi–Cd–Pb system, 96 °C for Bi–Pb–Sn, 104 °C for Bi–Cd–Sn, and 145 °C for Cd–Pb–Sn. They reported a quaternary eutectic at a composition 49.5 wt.% Bi, 10.1% Cd, 13.1% Sn, 27.3% Pb and a temperature of 70 °C. This quaternary eutectic composition is usually quoted as being close to that of Woods metal**, a fusible alloy containing 50% Bi, 12.5% Cd, 12.5% Sn and 25% Pb. In fact Lipowitz's alloy*** with 50% Bi, 10% Cd, 13% Sn and 27% Pb is pretty well on the quaternary eutectic composition.

A schematic representation of the polythermal projection of the Bi–Cd–Pb–Sn system is given in Fig. 273. This is broken down in Figs. 274–276 to allow separate representation of the

---

* N. Parravano and G. Sirovich, *Gazz. Chim. Ital.*, **42** (1) (1912) 630–716.
** Wood, *Dinglers Polytech. J.*, (1860) 158, 271.
*** W. C. Roberts-Austin, *An Introduction to the Study of Metallurgy*, Griffin & Co., London, 1891, p. 90.

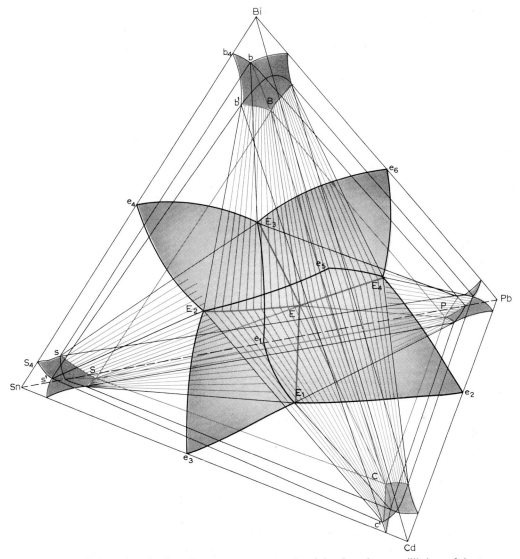

Fig. 273. Polythermal projection of a quaternary system involving five-phase equilibrium of the type
$l \rightleftharpoons \alpha+\beta+\gamma+\delta$ (schematic representation of the Bi–Cd–Pb–Sn quaternary eutectic system).

regions of primary, secondary and tertiary crystallisation respectively. In Figs. 275 and 276 the quaternary eutectic system has been represented in two ways; the first assumes complete insolubility of the components in the solid state and the second represents the more complicated state when solid solubility is introduced. There are W_i^n distinct spacial regions of i-ary crystallisation in an $(n-1)$-dimensional projection of an n-component phase diagram*, where $W_i^n = C_i^n$, the number of combinations of n objects taken i at a time. In Fig. 273 there are thus $C_1^4 = 4$ regions of primary crystallisation $(i = 1)$. These are shown in Fig. 274 as the primary Bi region $(Bi)e_6e_5e_4E_3E_4E_2$, the primary Pb region $(Pb)e_2e_1e_6E_4E_1E_3$, the primary Cd region $(Cd)e_2e_5e_3E_1E_4E_2$ and the primary Sn region $(Sn)e_3e_1e_4E_2E_1E_3$. There are $C_2^4 = 6$ regions

* A. I. LANDAU, *Zh. Fiz. Khim.*, **35** (1961) 2589; *Russ. J. Phys. Chem.*, (Engl. transl.), **35** (1961) 1279.

of secondary crystallisation in which the liquid phase is precipitating two solid phases simultaneously. Figure 275a shows these to be regions:

$$l \rightleftharpoons \text{Bi} + \text{Sn} \qquad \text{region (Bi)(Sn)}E_2EE_3$$
$$l \rightleftharpoons \text{Bi} + \text{Pb} \qquad \text{region (Bi)(Pb)}E_3EE_4$$
$$l \rightleftharpoons \text{Bi} + \text{Cd} \qquad \text{region (Bi)(Cd)}E_2EE_4$$
$$l \rightleftharpoons \text{Pb} + \text{Sn} \qquad \text{region (Pb)(Sn)}E_1EE_3$$
$$l \rightleftharpoons \text{Pb} + \text{Cd} \qquad \text{region (Pb)(Cd)}E_1EE_4$$
$$l \rightleftharpoons \text{Cd} + \text{Sn} \qquad \text{region (Cd)(Sn)}E_2EE_1.$$

Solid solubility (Fig. 275b) complicates matters. The $l \rightleftharpoons \text{Bi} + \text{Sn}$ region of secondary crystallisation for example is the space enclosed by the areas $b_4b^1E_2s^1s_4$ on face Bi–Sn–Cd, $b_4bE_3ss_4$

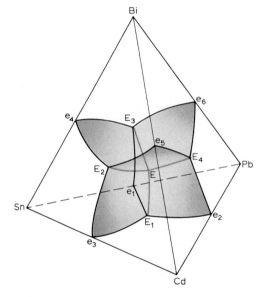

Fig. 274. Regions of primary crystallisation in the Bi–Cd–Pb–Sn system.

on face Bi–Pb–Sn and surfaces $bBEE_3b$, $b^1BEE_2b^1$, $s^1E_2ESs^1$, sE_3ESs, $b_4bBb^1b_4$ and $s_4sSs^1s_4$ in the tetrahedron. Referring to Fig. 273 separation of Bi and Sn from the melt occurs when the composition of the melt lies on surface $e_4E_3EE_2e_4$, the composition of the Bi solid solution on surface $b_4bBb^1b_4$ and the composition of the Sn solid solution on surface $s_4sSs^1s_4$ (compare with tie triangles $l_4\alpha_4\beta_4$ to $l_5\alpha_5\beta_5$ in the previous system (Fig. 268)).

There are $C_3^4 = 4$ regions of tertiary crystallisation, in which three solid phases are deposited simultaneously from the melt. Figure 276a indicates these phase regions to be:

$$l \rightleftharpoons \text{Bi} + \text{Sn} + \text{Pb} \qquad \text{region (Bi)(Sn)(Pb)}E$$
$$l \rightleftharpoons \text{Bi} + \text{Cd} + \text{Pb} \qquad \text{region (Bi)(Cd)(Pb)}E$$
$$l \rightleftharpoons \text{Bi} + \text{Cd} + \text{Sn} \qquad \text{region (Bi)(Cd)(Sn)}E$$
$$l \rightleftharpoons \text{Cd} + \text{Pb} + \text{Sn} \qquad \text{region (Cd)(Pb)(Sn)}E$$

Figure 276b shows that the regions of tertiary crystallisation are based on a tie triangle on a ternary face and extend into the quaternary as far as the tetrahedron formed by the four co-existing phases. For examples, the region for crystallisation of Bi + Cd + Sn from the melt

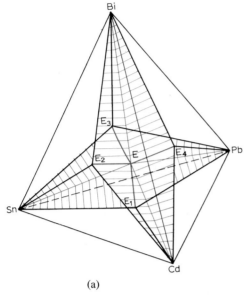

(a)

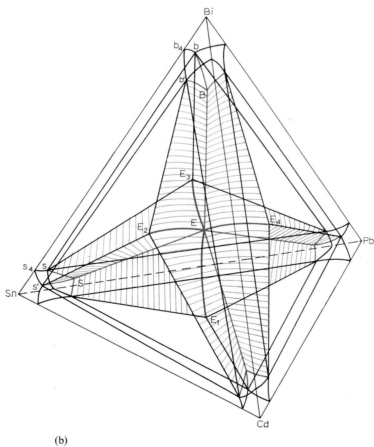

(b)

Fig. 275. Regions of secondary crystallisation in the Bi–Cd–Pb–Sn system, (a) assuming complete insolubility of the metals in the solid state, and (b) with solid solubility.

is represented by the space between tie triangle $b^1 s^1 c^1$ on the Bi–Cd–Sn ternary face and the tetrahedron $BSCE$. A series of tie tetrahedra are produced with apices on curves $b^1 B$, $s^1 S$, $c^1 C$ and $E_2 E$ for the deposition of Bi + Sn + Cd from the melt. The situation is identical with that shown in Fig. 268 for the crystallisation of $\alpha + \beta + \gamma$ from the tie tetrahedron $l_5 \alpha_5 \beta_5 \gamma_5$ to the tie tetrahedron $l_6 \alpha_6 \beta_6 \gamma_6$.

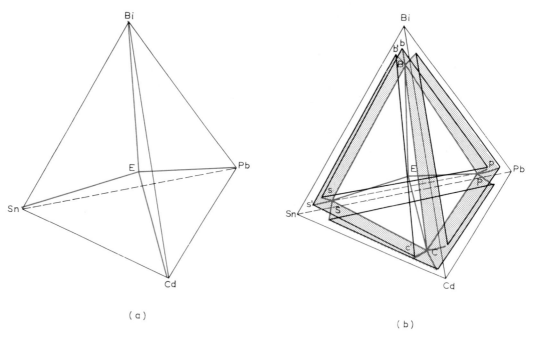

(a)   (b)

Fig. 276. Regions of tertiary crystallisation in the Bi–Cd–Pb–Sn system, (a) assuming complete insolubility of the metals in the solid state, and (b) with solid solubility.

It should be noted that a secondary crystallisation of Bi + Sn can be followed by either a tertiary crystlalisation of Bi + Sn + Cd or Bi + Sn + Pb. Which occurs is dependent on the course of the path of the liquid; it runs over surface $e_4 E_3 E E_2 e_4$ during separation of Bi + Sn and, if it meets curve $E_2 E$, equilibrium will be established between Bi (along curve $b^1 B$), Sn (along curve $s^1 S$) and Cd (along curve $c^1 C$). If, however, the liquid meets curve $E_3 E$, equilibrium will be established with Bi (curve bB), Sn (curve sS) and Pb (curve pP).

Finally, there is one region of quaternary crystallisation where the invariant quaternary eutectic reaction occurs with isothermal separation of Bi + Pb + Sn + Cd from the remaining melt. The melt has a composition represented by point E (Fig. 273) and the four solid phases have compositions B, P, S and C. Solidification is therefore completed when the liquid composition, which is tracing a path along one of the curves $E_1 E$, $E_2 E$, $E_3 E$ or $E_4 E$ reaches the quaternary eutectic point E.

Figure 277a indicates the path traced by the liquid of initial composition X within the primary Bi field. The Bi solid solution would trace the path shown in Fig. 277b: from point 1 to 2 primary Bi is deposited, from 2 to 3 secondary Bi, from 3 to B tertiary Bi and at point B the quaternary eutectic Bi separation. The paths of the Sn and Cd solid solutions can also be plotted provided that it is remembered that the Sn solid solution will trace a curve of the type 2–3–S and the Cd solid solution a curve of the type 3–C, since Sn is not precipitated from the liquid

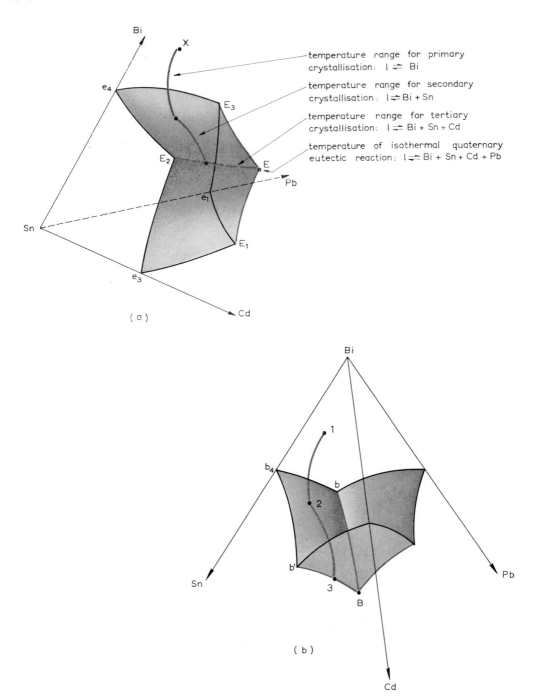

Fig. 277. (a) Change in liquid composition during freezing of an alloy whose composition lies in the primary Bi phase region, (b) corresponding change in composition of the Bi solid solution.

until the secondary crystallisation begins and Cd does not separate from the liquid until the tertiary crystallisation begins. The Pb solid solution does not trace a path at all since it appears only at the quaternary eutectic temperature when its composition is fixed at point *P*.

15.6. TWO-DIMENSIONAL POLYTHERMAL SECTIONS

The serious student should refer to Palatnik and Landau's work* for methods applicable to the construction of two-dimensional polythermal sections of quaternary systems. Most use is made of the simple quaternary eutectic system with complete immiscibility assumed in the solid state, although mention is made of methods for tackling the usual case with partial solubility in the solid state.

15.7. COMPLEX QUATERNARY SYSTEMS

A theoretical study of more complex quaternary systems has been made by Cayron** and the constitution of eight Al-base quaternary systems has been surveyed by Phillips***. The latter work includes reference to the reasons for failure to reach equilibrium under commercial conditions of casting and fabrication.

  * L. S. PALATNIK AND A. I. LANDAU, *Fazovye Ravnovesiya v Mnogokomponentnykh Sistemakh* (Phase Equilibbria in Multicomponent Systems), Khar'kov State University, Kar'kov, 1961, pp. 270–320.
 ** R. CAYRON, *Etude Théorique des Diagrammes d'Equilibre dans les Systémes Quaternaires*, Institut de Metallurgie, Louvain, 1960.
*** *Equilibrium Diagrams of Aluminium Alloy Systems*, Aluminium Development Association Inform. Bull. No. 25, London, 1961.

Selected Bibliography

GENERAL

W. HUME-ROTHERY AND G. V. RAYNOR, *The Structure of Metals and Alloys*, Monograph No. 1, Institute of Metals, London, 4th edn., 1962. (A more elementary treatment of the same subject is W. HUME-ROTHERY, *Elements of Structural Metallurgy*, Institute of Metals, London, 1961.)

A. H. COTTRELL, *Theoretical Structural Metallurgy*, Edward Arnold, London, 2nd edn., 1955.

W. C. WINEGARD, *An Introduction to the Solidification of Metals*, Institute of Metals, London, 1964.

R. A. SWALIN, *Thermodynamics of Solids*, Wiley, New York, 1962.

L. S. DARKEN AND R. W. GURRY, *Physical Chemistry of Metals*, McGraw-Hill, New York and London, 1953.

J. D. FAST, *Entropy*, (translated from the Dutch by Mrs. M. E. Mulder-Woolcock, Eindhoven), Cleaver-Hume Press, London, 1962.

E. A. GUGGENHEIM, *Thermodynamics*, North-Holland Publ. Co., Amsterdam, 1949.

The Scientific Papers of J. Willard Gibbs, Vol. 1, *Thermodynamics**, Dover Publications, New York, 1961.

PHASE EQUILIBRIA

M. HANSEN AND K. ANDERKO, *The Constitution of Binary Alloys*, McGraw-Hill, New York, 1958.

J. L. HAUGHTON AND A. PRINCE, *The Constitutional Diagrams of Alloys: A Bibliography*, Institute of Metals, London, 2nd edn., 1956.

G. MASING, *Ternary Systems*, (translated from the German by B. A. Rogers), Reinhold, New York, 1944.

F. N. RHINES, *Phase Diagrams in Metallurgy*, McGraw-Hill, New York, 1956.

R. VOGEL, *Die Heterogenen Gleichgewichte*, Geest & Portig, Leipzig, 2nd edn., 1959.

R. CAYRON, *Etude Théorique des Diagrammes d'Equilibre dans les Systémes Quaternaires*, Vols. I and II, Institut de Metallurgie, Louvain, 1960.

L. S. PALATNIK AND A. I. LANDAU, *Phase Equilibria in Multicomponent Systems*, (translation by J. Joffe of the Russian edition published by Khar'kov University, 1961), Holt, Rinehart and Winston, New York, 1964.

JOURNALS

Journal of the Institute of Metals.

Journal of the Less-Common Metals.

Transactions of the Metallurgical Society of the American Institution of Mining and Metallurgical Engineers.

Zeitschrift für Metallkunde.

Zhurnal Neorganicheskoi Khimii (available in translated form as *Russian Journal of Inorganic Chemistry*).

---

* The symbols used by Gibbs for entropy, energy, heat content, Helmholtz free energy and Gibbs free energy were η, ε, χ, ψ and ζ, corresponding with the symbols S, E, H, F and G used in this book.

Index